Edwin J. Houston, Arthur E. Kennelly

Electricity Made Easy

By simple language and copious illustration

Edwin J. Houston, Arthur E. Kennelly

Electricity Made Easy
By simple language and copious illustration

ISBN/EAN: 9783337393922

Printed in Europe, USA, Canada, Australia, Japan

Cover: Foto ©berggeist007 / pixelio.de

More available books at **www.hansebooks.com**

ELECTRICITY MADE EASY

BY

SIMPLE LANGUAGE AND COPIOUS ILLUSTRATION

BY

EDWIN J. HOUSTON, Ph. D. (Princeton)

AND

ARTHUR E. KENNELLY, Sc. D.

NEW YORK

AMERICAN TECHNICAL BOOK COMPANY

45 Vesey Street

1898

TABLE OF CONTENTS.

ELECTRICITY MADE EASY

BY SIMPLE LANGUAGE AND COPIOUS ILLUSTRATION.

CHAPTER I.

THE TURNING-ON OF AN ELECTRIC LAMP IN THE HOUSE.

EVERY one knows that when a spigot connected with one of the water faucets in a house is turned, as, for example, the spigot S, at the wash-stand shown in Fig. 1, the water runs out of the pipe, at the faucet F, and will continue running out as long as the spigot is left open. When the spigot is opened, we say that the water is *turned on*; when it is closed, we say that the water is *turned off*.

The water flows out of the pipe, as soon as an opening is made by the turning of

Fig. 1. Turning on water at wash-stand.

the spigot, because the water is constantly pressing against the inside of the pipe. When no opening exists, the water simply presses against the pipe, but does not run out until an opening is made.

We can both see and feel the water running out of the pipe. We can fill a tumbler or other vessel with the water; we can drink the water, or can use it for washing, cooking and other well-known purposes. We introduce water into the house in order to make use of it for the many purposes for which it is adapted.

W W' K F B

Fig. 2. Turning-on light at incandescent electric lamp.

Every one, too, knows that when we turn the key connected with an incandescent electric lamp, as, for example, that shown at K, in Fig. 2, that the lamp glows, or throws out light, and will continue throwing out light as long as the electricity continues to flow through the filament. When the key is again turned, the *electric flow*, or *current*, is stopped, and the lamp ceases to throw out light. The lamp glows, or throws out light, because electricity flows from the wires *W*, *W'*, through the slender carbon thread or filament F. When the key K, is turned along the lamp, as shown in the figure, so as to permit the electricity to flow, we say that the electricity is *turned on;* when the key is turned across the lamp, so as to prevent the electricity from flowing, we say that the electricity is *turned off.*

The electricity comes out of the wires *W*, *W'*, connected with the lamp, and flows through the lamp filament, as soon as a conducting path is provided for it by turning the key, and will continue flowing through the lamp as long as the electricity is turned on. The electricity is turned off, by the turning of the key, because the conducting path through the lamp is broken.

The electricity flows out of the supply wires *W*, *W'*, and passes through the lamp, as soon as a path is provided for it, because the electricity is constantly exerting a pressure upon the wires and is endeavoring to escape from them. When the key is so turned as to cut off the path between the supply wires and the lamp, the electricity continues to exert this pressure, and is ready to escape at any time from the wires that a path is offered it.

Let us now examine how the turning of the spigot of a water faucet, permits the water to flow from the pipe through the faucet. A simple mechanism for doing this is shown in Fig. 3, which represents the water as turned on at A, and turned off at B. Here a tightly fitting metallic plug P, provided with an opening S S, extending completely through it, is turned by the movement of the spigot so as to have this opening extend in the direction of the length of the pipe, when the water is turned on, and across the length of the pipe, when the water is turned off. It is

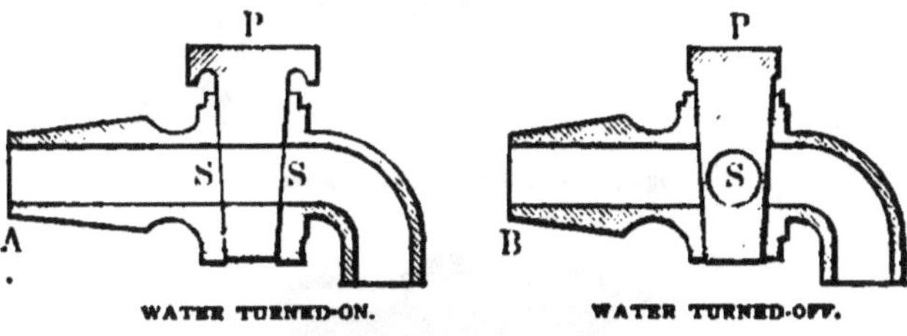

Fig. 3. Mechanism of water faucet and spigot.

evident that the water stops flowing, when the plug is in the position shown at B, because the path or opening through the pipe is then cut off. A very common form of water faucet or *bibb*, is shown in Figs. 4 and 5. Here the turning of the spigot S, moves a screw S′, carrying a valve V. The movement of S, opens or closes the

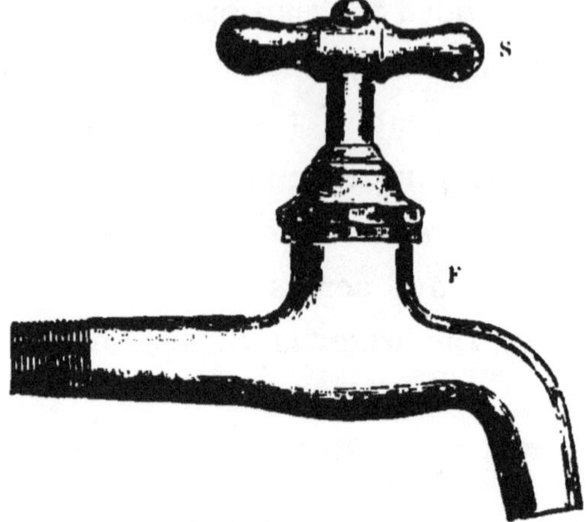

Fig. 4. Valve form of water faucet.

Fig. 5. Details of valve of water faucet. Water turned on.

faucet by raising or lowering the valve. The position of the valve shown in Fig. 5, is that in which the water is *turned on.*

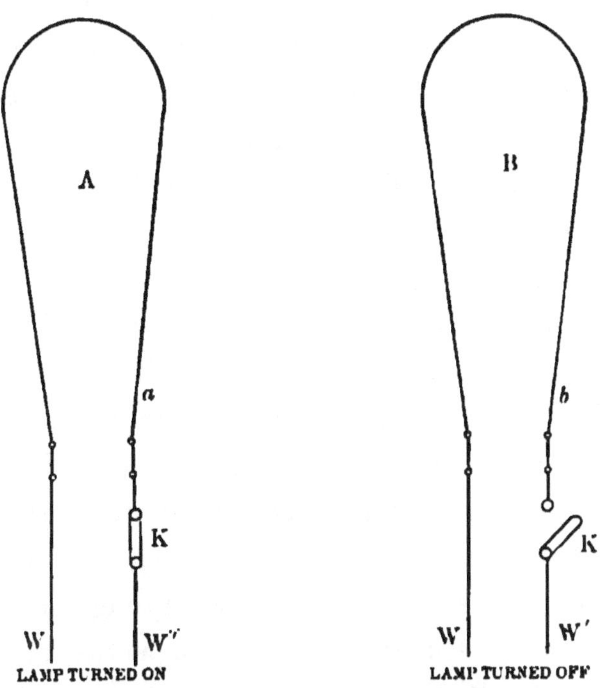

Fig. 6. Diagram of incandescent lamp and key.

The manner in which the turning of the key of an electric lamp permits the electricity to flow through the lamp is shown diagrammatically in Fig. 6, where the lamp is represented as turned-on at A, and turned-off at B. Here turning the key K, so as to place it in the position shown at A, completes the conducting path between the supply wires *W*, *W'*, and the ends or terminals of the lamp filament, and permits the electricity to leave the wires, and pass through the lamp, while turning the key so as to leave it as shown at B, produces a break in the conducting path and prevents the electricity from flowing through the lamp. The electricity cannot pass across the air space left between b and K, because the air in the opening is *non-conducting*; i. e., will not permit the electricity to flow through it.

When the gas is turned on at the gas burner B, Fig. 7, by turning the key K, the gas flows out from the pipe, and will continue flowing out as long as the gas remains turned on. The turning of the key establishes an opening between the pipe and the burner, by means similar to the opening established in a water pipe by the turning of the spigot. An examina-

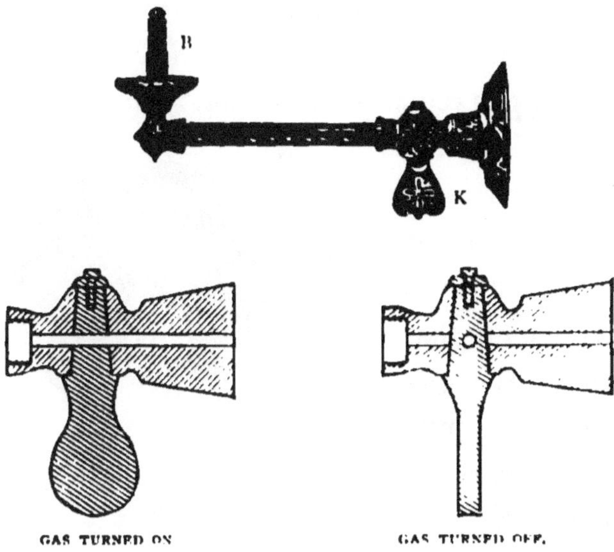

Fig. 7. Section of gas-burner key.

tion of Fig. 7, will show that when the gas is turned on, the key is moved so that an opening through the key plug extends in the direction of the length of the pipe, thus permitting the gas to flow through the plug, and that when the gas is turned off, the key plug is turned so that the opening is at right angles to the length of the pipe.

If the escaping gas is lighted it will burn and give off light and heat as long as the gas remains turned on. Sometimes an electrical device, such as is shown in Fig. 8, is employed, whereby the pulling of a pendant chain *p*, both turns the gas on and causes a minute electric spark to jump through the issuing gas stream, thus igniting it. In such cases the ordinary thumb key *k*, is left open, and an additional key is provided, operated by the movement of the pendant *p*. Here, one pull of the pendant turns the gas on and lights it, and the next pull of the pen-

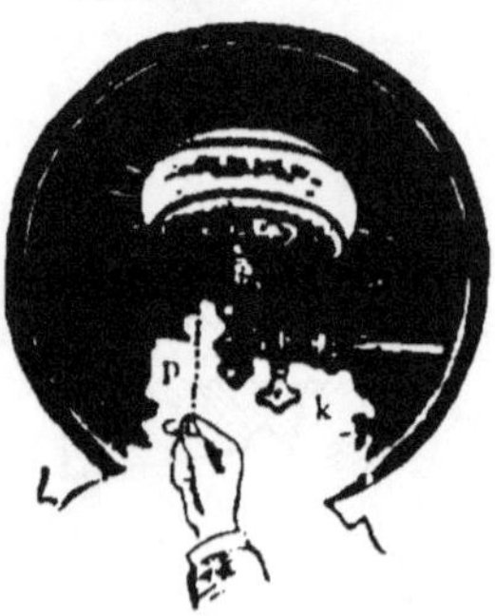

Fig. 8. Device for both turning-on and lighting a gas jet.

dant turns the gas off, and extinguishes the light. The manner in which the electrical part of this device operates will be explained in a subsequent chapter.

We have shown diagrammatically in Fig. 6, how the electricity is turned on at an electric lamp by bridging over an air gap between the supply wires and the lamp terminals. Let us now inquire how this is done, and how the lamp is connected to the supply wires or conductors. Fig. 9, shows a lamp ready for connection with the supply wires. The lower part of the lamp is called the *lamp base.* The base is furnished with a metallic thread *b*, connected with one of the terminals of the lamp filament, and a metallic piece *a*, insulated from *b*, and connected with the other terminal of the lamp filament.

Fig. 10, represents a form of *lamp socket*, suitable for a lamp such as shown in Fig. 9. The lamp socket is the name given to

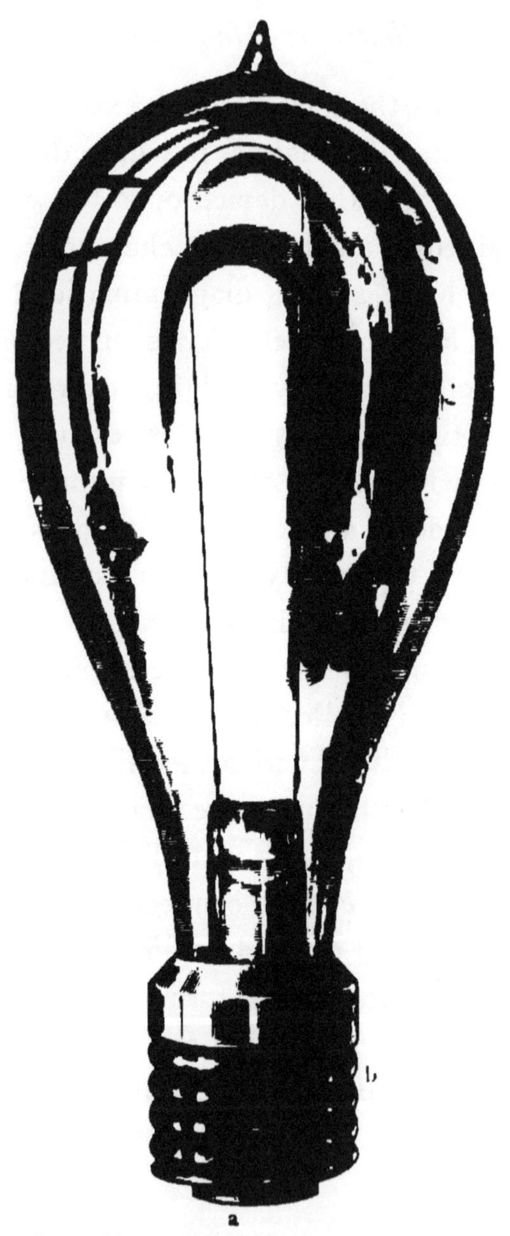

Fig. 9. Lamp with screw threaded base for attachment to lamp socket.

the device provided for holding the lamp base. The lamp socket contains the metallic contact pieces connected to the ends of the supply wires that bring the electric current to the lamp, and is so arranged that the mere insertion of the lamp into its socket electrically connects the lamp with these pieces, and thus with the supply wires. One of the supply wires is connected to the metallic thread *b'*, and the other to a metallic piece not shown in the figure, but shown in Fig. 11, at *a'*. When the lamp is screwed into its socket, the thread *b*, Fig. 9, brings one of the lamp terminals into electric contact with one of the lamp supply wires

Fig. 10. Lamp socket showing one of screw connections.

Fig. 11. Lamp base showing both screw connections.

through b′, Fig. 11, and the piece *a*, Fig. 9, brings the other lamp terminal into contact with the other supply wire through a′, Fig. 11.

We will now examine how the turning of the key K, is able to open and close the connections between the lamp and the supply wires. To do this let us examine Fig. 12, which shows an Edison incandescent electric lamp with a section through the lamp base and socket. The supply wires are shown at *W*, *W′*. S, is the lamp socket and B, the lamp base, with the lamp terminals connected to *A* and *B*, and the supply terminals to *A′* and *B′*. An inspection of the figure will show that when the lamp base is screwed down into its socket, *A*, comes into contact with *A′*, and *B*, in contact with *B′*, thus connecting the lamp terminals with the supply wires. When the key K, is in the position shown in Fig. 10, the current flows through the

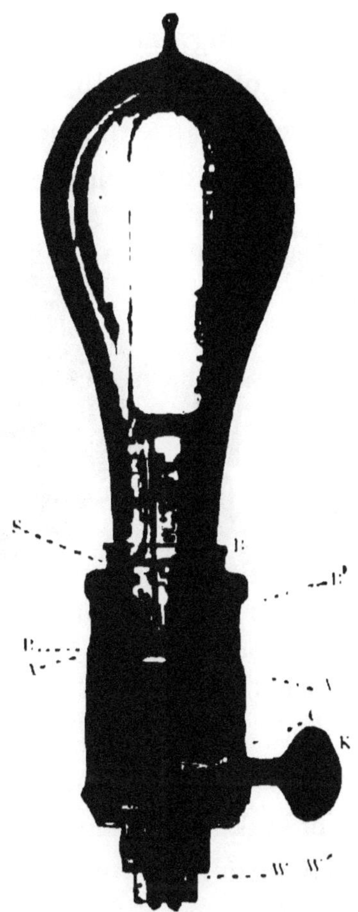

Fig. 12. Edison lamp with section through the socket.

lamp because the contact piece *C*, touches a contact spring, but when K, is moved through a quarter turn, the current no longer flows, because the piece *C*, is turned so as no longer to touch the spring.

Fig. 13. Key socket. Key in position for turning electricity on.

The forms of lamp-socket shown in Figs. 10, 11 and 12, are called *key-sockets*, because the lamp can be turned on or off by means of a key K, placed in the socket. A similar form of key socket is shown in Fig. 13. Contact pieces, connected with the supply wires, are so arranged that when the lamp base is screwed into position in the socket, one lamp terminal is connected with one electric supply wire at *a*, and the other lamp terminal with the other supply wire by the metallic spring at *s*, by the piece *b*, coming into contact with it when the key K, is in the position shown in the figure; but when the key is turned through a quarter turn, the current is turned off, because contact is broken between *b* and *s*. A cover S′, is placed over the contacts, etc., to protect them from dust.

Sometimes it is found more convenient to turn the lamp on by means of a *lamp key* or *switch* at a distance. This is especially the case when a number of lamps are to be turned on or off at the same time, as in an *electrolier*, the name given to an electric chandelier. Here a form of socket called a *keyless socket* is used. In keyless sockets a slight rotation will extinguish a burning lamp by *breaking* or disconnecting one of the contacts between the terminals of the supply wires in the socket, and the lamp terminals in the lamp base. A form of keyless socket is shown in Fig.

Fig. 14. Keyless socket.

14. When the lamp is screwed in position, contacts between the lamp terminals and the supply terminals are made at *a* and *b*.

In what are called *combination fixtures*, provision is made for using electricity, or gas, or both. A combination fixture, for a single incandescent lamp and gas burner, is shown in Fig. 15.

The turning-on of electricity, so as to permit it to flow through any device placed in its pathway, may be attended by

Fig. 15. Combination fixture for both gas and electricity.

other effects than the lighting of a lamp. These will be discussed in subsequent chapters. Reference, however, may here be made to the power of electricity to operate an electric motor for driving a fan, as in the *ceiling fan* shown in Fig. 16.

Here an electric motor is placed inside the metal cover as shown, and directly coupled

Fig. 16. Electric ceiling fan.

to the fan. The wires carrying the current into and out of the motor pass from the ceiling through a pipe, W. Or, the electricity may flow both through a number of lamps and also through an electric

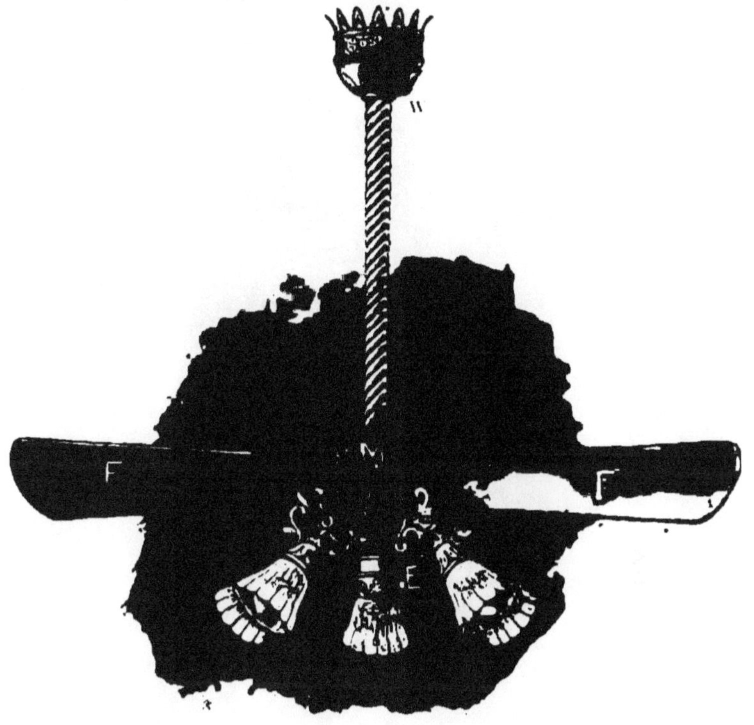

Fig. 17. Electrolier and ceiling fan.

motor, so as to operate the combined electrolier E, and the fan F F, shown in Fig. 17.

A key or switch K, suitable for turning on or off a number of lamps at one time, is shown in Fig. 18. It is called a *flush switch*, because its key is flush with, or

Fig. 18. Flush switch.

does not project beyond, the surface of the wall in which it is placed.

The gas escapes from a gas-pipe on the opening of the key, because the gas exerts a pressure against the inside of the pipe and is trying to escape from it. If an opening is made in the pipe, the gas will run out. If even a small crack or fissure exists in the pipe some gas will escape, or flow, and thus be lost.

We cannot see the gas escaping, since it is invisible. In this respect it differs from the water. We can, however, smell it. We employ it for various purposes: namely, for lighting and heating our houses.

Electricity like gas is invisible. If we let it pass through our bodies, it produces various sensations and may even give so severe a shock as to cause death. Electricity differs from gas or water in that it is not a material or gross substance. Unlike either gas or water it has no weight. It flows through wires very much like water or gas flows through pipes, but flows through a solid wire as readily as through a tubular wire of the same weight.

We introduce electricity into a building, in order to utilize the many useful properties which it possesses. We can light and heat buildings, drive elevators, motors and fans, send telegrams, and telephone messages to distant cities, and do many other things with it. In most cases it is not at all difficult to know how electricity serves us in these different respects, and we will try in this book to aid you in understanding this.

The real nature of electricity is still unknown ; but, while we are thus ignorant of its nature, we are well acquainted with the laws or conditions under which it operates. Indeed, the electrical engineer is, perhaps, better acquainted with the laws of electricity than the mechanical engineer is with the laws of mechanics. Many of the laws of electricity are exceedingly easy to understand, if you are willing to give them a little thought. We are going to see if we cannot help you to understand some of these laws.

CHAPTER II.

HOW THE ELECTRIC WIRES ARE DISTRIBUTED THROUGH THE HOUSE.

IF we could trace the pipe F, Fig. 19, connected with the water faucet A, say in a bath-room on the third floor of a house, we would find that it extends down through the floors beneath, into the cellar, from whence it passes, generally under the pavement, to one of the water-pipes or *mains* M, in the streets. These street mains are kept filled with water under pressure, and it is this pressure which causes the water to rise from the street mains and enter the pipes of the house.

In the modern house there are generally many water-pipes. For example, there will be the pipe leading to the faucet in the front of the house, provided for convenience in washing the pavement;

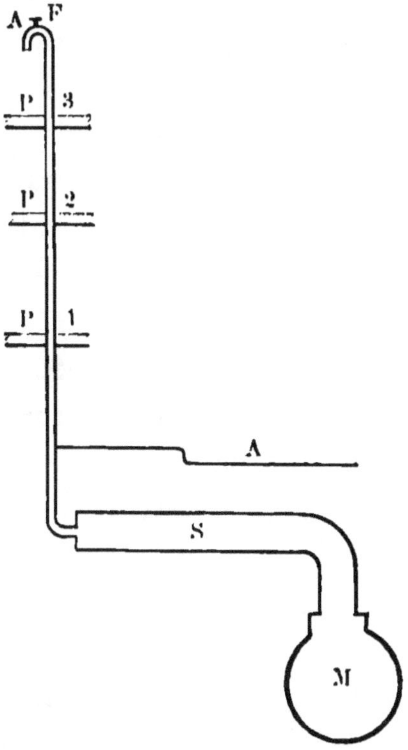

Fig. 19. Carrying capacity of water pipes.

then there is the pipe which leads to the hydrant in the yard ; other water-pipes furnish the water that flushes the basins, still others lead to and from the kitchen sink, the range, the wash-tubs, the pantry, the bath-rooms and the stationary wash-stands of the house. In a large house these make a network of pipes. As a rule, a number of faucets are connected with a single pipe, and all the pipes of a house are connected with a single service pipe S, Fig. 19, that joins the water main M, in the street.

It will be noticed that the service pipe is much larger than the other pipes in the house. This is because all the water used in the house passes through this pipe. For a similar reason, the street mains, which supply all the houses in a street with water, are necessarily much larger than the service pipes of the separate houses.

Generally, the size of the various water pipes that lead to different parts of the house will vary with the quantity of water that has to pass through them. When only a single faucet has to be supplied, a comparatively small pipe will answer, but when a number have to be supplied from a single pipe, a larger pipe is necessary. Thus, the water pipe shown at A, Fig. 20, where only a single faucet is supplied, is a smaller pipe than that shown at B, where three faucets are supplied. It is true that where only a single faucet is apt to be used at any one time, a smaller pipe will answer, but when all three faucets are discharging at the same time, it is evident that the *carrying capacity* of B, must be three times as great, or that its diameter or width must be greater than that of A.

If the gas pipes of the house be examined in the same manner, it will be noticed that they, too, form a network of pipes. The service main, which enters the house from the street, is larger than any of the other pipes, the pipes growing

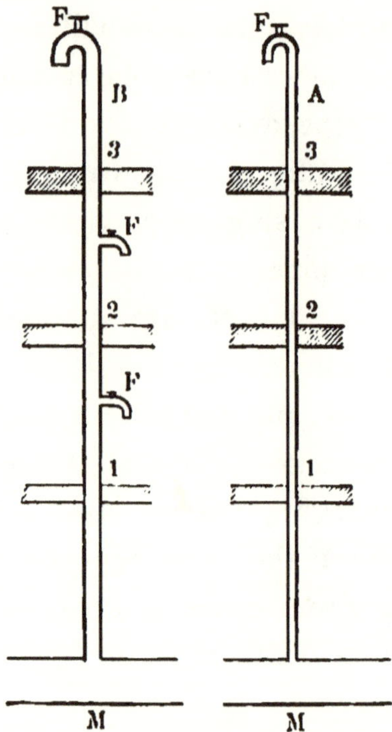

Fig. 20. Carrying capacity of water pipes.

smaller as they pass into the branches, and sub-branches that supply fewer and fewer lights. A small pipe will suffice for a single burner, but a large pipe must be supplied to furnish the gas necessary for a large chandelier.

The electric light wires which supply electricity to different parts of a large house form a network of wires. If we could trace these wires we would find, as in the case of the gas or water pipes, that they have a common connection with large wires or *electric mains* in the street. In the case of the electric wires there is, however, this difference; that whereas, a single water or gas pipe suffices for the introduction of water or gas into a building, two wires are necessary for the introduction of electricity; one for the electricity to leave one of the mains in the street and enter the house, and another for it to return to the other main in the street, after it has passed through the lamps and other devices in the house.

Tracing the electric wires as they enter

the house from the street, we first find comparatively large and heavy sets of copper wire called the *service wires.* Since all the electricity which is used in the house must pass through the service wires, their size will necessarily depend on the number of lamps or other devices that are to be operated by the current. If their size be too small, there might be supplied to the lamps, or other devices in the house, a quantity of electricity insufficient to properly operate them, so that the lamps would burn too dimly, and the fan motors revolve too slowly; but, here also, as in the case of water or gas service, if only a part of the lamps are turned on, no defect will be noticed in their operation; since, under these circumstances, sufficient electricity can pass into the house through the service wires to supply the lamps that are in use.

In a fairly large house, where a great

number of lamps are to be supplied, such as that shown in Fig. 21, pairs of conductors called *risers* R, R, pass from the serv-

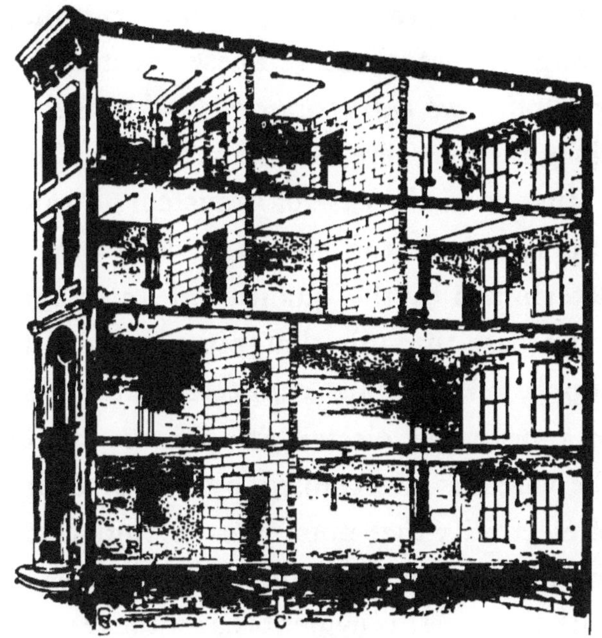

Fig. 21. Plan showing wiring of house.

ice wires up to the different floors. The risers are connected with a number of pairs of smaller wires which are carried to the lamps.

In general, the risers extend from the service wires in the cellar to the various corridors of the house, where they connect with wires called *floor-mains* running along the corridors. These floor-mains give off "*sub mains* or *branches*" to the various rooms where the lamps are installed.

Just as in the case of a system of water supply, where the size of a water pipe is determined by the quantity of water that must pass through that pipe in order to supply all the taps connected with it, so in a system of electric supply, the size of the wires is determined by the quantity of electricity that must pass through them in order to supply all the lamps connected with them. It is evident that if the *electric carrying* or *conducting capacity* of a system of electric wires is unnecessarily great, money will be needlessly spent in putting in heavier wires than are required, and, if the conducting capacity is too small, the lamps will fail to operate properly, when all are turned-on, because the wires are unable to supply sufficient current.

Where a single lamp A, is to be sup-

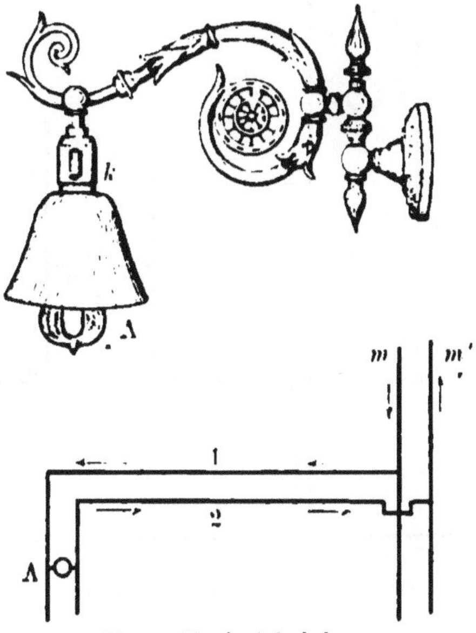

Fig. 22. Circuit of single lamp.

plied, the wires may be arranged as shown in Fig. 22, where a pair of wires 1, 2, pass from the sub-mains *m*, *m'* to the lamp at A. Here it will be seen that the electricity, which for convenience is assumed to pass from the sub-main *m*, to the *branch* 1, can only return through the branch 2,

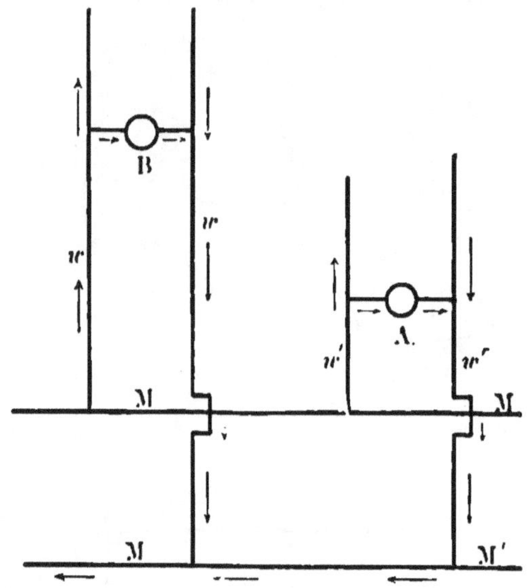

Fig. 23. Separate circuits for two lamps.

and *m'*, by passing through the lamp at A.

When two lamps *A* and *B*, Fig. 23, are to be fed or supplied from a single pair of wires *M*, *M*,' two separate pairs of wires *w*, *w*, and *w′*, *w′*, may be run from the sub-mains as shown in the figure; or, as is more convenient, and is practically always done, a single pair of wires *w′*, *w′*, Fig. 24,

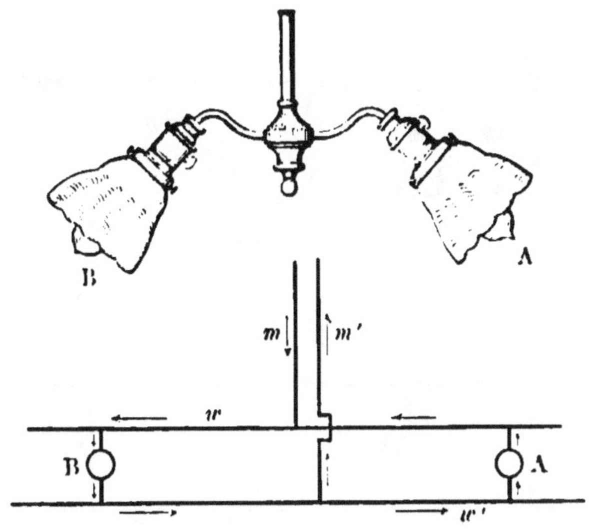

Fig. 24. Connection of two multiple-connected lamps.

having twice the carrying capacity of the separate pairs of Fig. 23, are employed and the lamps connected as shown. In a similar manner the connections of four lamps 1, 2, 3 and 4, are shown in Fig. 25, and of six lamps 1, 2, 3, 4, 5 and 6, in Fig. 26, between *m*, *m'*, and *w*, *w'* are as therein indicated.

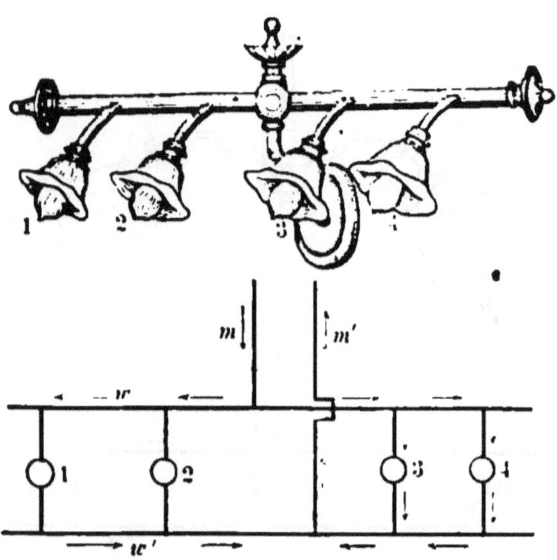

Fig. 25. Circuit of multiple connections of four lamps.

The connection of a number of separate lamps to a single pair of wires, as shown in Figs. 24, 25 and 26, is similar to the connection of a number of separate faucets to a single water pipe, as for example, the three faucets shown in Fig. 20. In this latter case of water, a single pipe only is necessary, unless, indeed, the waste pipe, that carries the water to the sewers, be re-

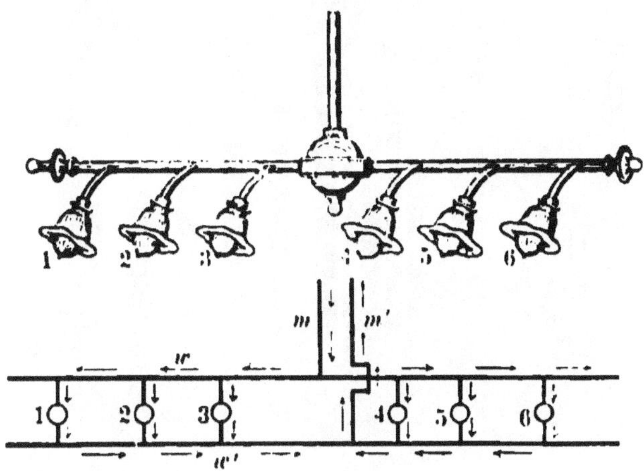

Fig. 26. Circuit connections of six multiple-connected Lamps.

garded as the return conductor. But in the case of the electric supply, two wires are necessary, one by means of which the electricity enters the lamp from one main, and the other by means of which it leaves the lamp and returns to the other main.

The connection of a number of electric

Fig. 27. Multiple connection of pipes in steam radiator.

lamps to a single pair of wires, as shown in Figs. 24, 25 and 26, is called connection in *multiple*, or in *parallel*. It is similar to the connection often made in the steam or hot-water radiators employed in heating a building. A steam radiator is shown in Fig. 27, where the steam entering through

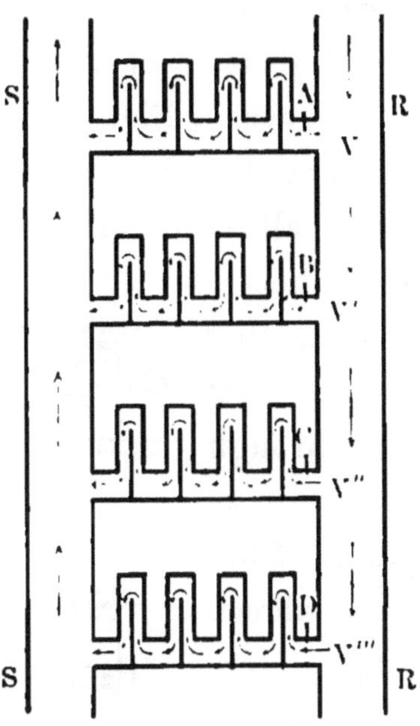

Fig. 28. Diagram showing four multiple-connected radiators.

a large pipe controlled at the valve V, by a wheel handle W. Fig. 28, shows, diagrammatically, four steam radiators A, B, C and D, connected in multiple, between the pipe R R, that supplies them with steam, and the pipe S S, that returns the hot water and steam to the boiler. Sometimes, however, when the pressure is sufficiently great, instead of arranging the radiators, as shown in Fig. 28, so that the steam or hot water divides and passes through them all in parallel or abreast, they are so connected to one another that the steam or hot water passes through each successively, as in the system of hot water heating, shown in Fig. 29. Here two radiators, B and C, have their pipes so connected to the boilers, and to one another, that the hot water leaving the

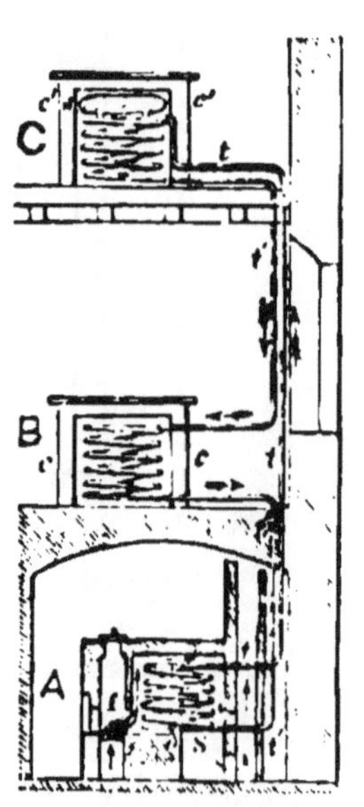

Fig. 29. Series connection of hot water radiators.

boiler A, passes first through C, then through B, and finally back through the boiler. Such a connection is called a *series connection.* In a series connection any stoppage at one point in the pipe would cut off the hot water from the remaining radiators unless a bypath was provided for its passage around the radiator thus cut off.

With the multiple connection of pipes or wires this is not the case. Any single device, such as a radiator or a lamp, can be cut-off without affecting the remaining radiators or lamps. Thus, in Fig. 28, which shows the multiple connection of four radiators A, B, C and D, in order to cut out A, it is only necessary to close its valve V; or, similarly, to cut out B or C, to close the valve V′, or V″. In the same manner it is evident that in the multiple-connected lamps, shown in Figs. 24, 25 and 26, any lamp can be cut out by dis-

connecting one of its wires, without influencing the remaining lamps.

The series connection is frequently employed in the supply of electricity to a

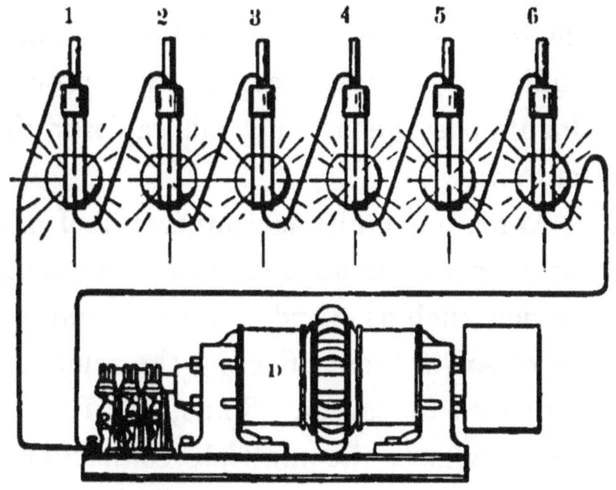

Fig. 30. Series connection of six arc lamps.

number of arc lamps for street lighting. Instead of dividing the wires into branches and placing the lamps in each branch, the separate lamps are placed in breaks in the wire as shown in Fig. 30, where six

arc lamps 1, 2, 3, 4, 5 and 6, are connected in series, with the circuit of the dynamo D. Here the electricity passes say first through the lamp 1, and after leaving this lamp flows through the circuit to lamp No. 2, and thence successively to Nos. 3, 4, 5 and 6. This connection is called *series connection.* The multiple connection is generally employed for the distribution of the small or incandescent lamps, and the series connection, for the distribution of the large or arc lamps. Arc-lamp circuits are sometimes over twenty miles in length, the electricity passing successively through more than one hundred arc lamps. In such cases the electric pressure required to drive the electricity through all the lamps is great.

It is sometimes convenient to combine the multiple and series connections. For example, the six incandescent lamps shown in Fig. 31, may be connected, as shown, in three parallel groups of two each, so that the electricity leaving A, divides into three branches flowing in

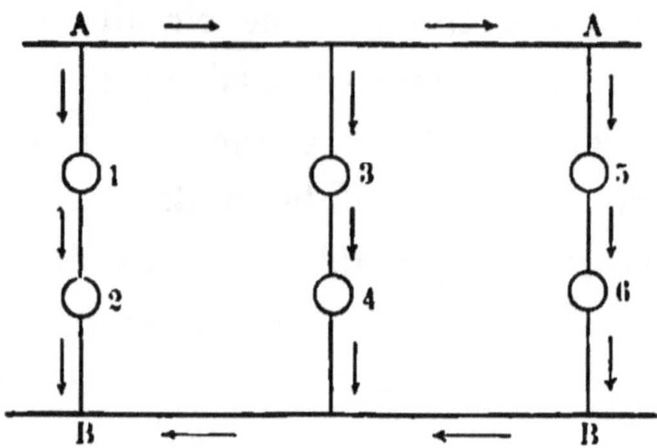

Fig. 31. Multiple-series connection of six lamps.

multiple or parallel through the two series-connected lamps in each of the three branches.

CHAPTER III.

HOW THE ELECTRIC STREET MAINS SUPPLY THE HOUSE.

WHILE the water is running out of the pipes through the various faucets that may be opened in a house, the pipes are kept filled by the water in the street mains under the influence of the pressure acting thereon. Let us now examine how in their turn these mains are kept filled with water. If we follow the water mains, in imagination, through the streets of any city, we will find that they all converge or lead towards one or more points of supply. These points generally consist of what are called *reservoirs*. As we near the reservoir, the water-pipes increase in size, those directly connected with the reservoir being the largest in the distribution system, as shown in Fig. 32. The reservoir is generally kept filled with water by means of pumps driven either by steam or by water power.

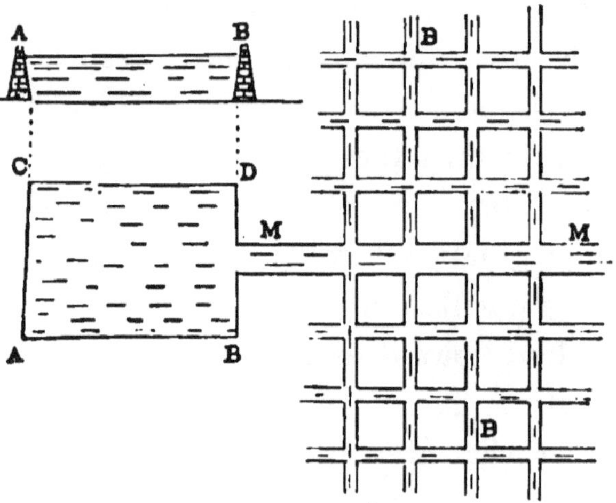

Fig. 32. System of water mains and their connection to reservoir.

The reservoir is situated on a hill, or other elevation, and the pipes which enter near the bottom are kept filled with water under pressure by reason of the pressure of the water. The amount of this water pressure, at any point in a pipe, depends on the height of the water in the reservoir above that point, and the situation of the pipe as regards the reservoir. As the water runs out of the reservoir into the pipes, the water is kept at the same level in the reservoir by the action of the pumps. The water that escapes from the faucets in the house, therefore, comes from the water in the reservoir, the pipes merely serving as a path between the reservoir and the places in the house where the water escapes.

In the same manner, if we follow in the imagination the gas-pipes from a house through the streets of a city, we will find them leading to a large reservoir of gas in the shape of a *gas-holder* H, such as that shown in Fig. 33. Gas-holders generally consist of two separate cylinders, the upper sliding or telescoping into the lower. A portion of the weight of the holder exerts a pressure on the gas in the gas main, and it is this pressure which causes the gas to escape at the gas-burners when they are opened. As the gas flows out of the gas-

Fig. 33. Gas-holder or reservoir for illuminating gas.

holder into the pipes, and through the pipes out of the various gas-jets, the holder is kept filled with fresh gas produced by the action of heat on bituminous coal, or by other suitable means.

So too, in the case of the steam in the steam-pipes connected with the radiators

Fig. 34. Steam heater showing fire box and steam heating pipes.

employed in heating the house. If we follow them we will find them terminating in a boiler where the steam is produced by the burning of coal, as shown in Fig. 34, which represents a steam heater. It is the pressure of the steam that causes the steam to flow from the boiler into the pipes. The steam or hot water returns to the boiler, after having parted with some of its heat in the radiators. As the steam passes out of the boiler into the pipes, the boiler is kept filled with fresh steam by the action of the heat on the water in the boiler, which is kept filled with water by any suitable means.

If in like manner we follow in imagination the electric mains through the streets of a city, we will find them all converging to a place called a *central station*, where the electricity is produced.

But before we enter the central station it may be well to examine in some little detail the manner in which the electric wires or conductors are placed under the streets. In general, there are two ways in which this can be done; viz., by the use of *electric conduits*, and by the use of *electric tubes.* An electric conduit consists of an underground passage-way, or space, provided for the reception of electric cables or conductors. Conduits may be formed of tubes of earthenware or iron; or of creosoted or tarred wood. The conductors, besides being covered with some insulating material, are also protected by a covering or sheathing of lead.

A form of wooden conduit is shown in Fig. 35. It consists of wooden pieces placed together as shown. The wood is covered with hot pitch, or is impregnated with some chemical preservative such as creosote, to preserve it from rapid decay. A form of glazed stone-ware conduit, containing 25 ducts, is shown in Fig. 36. The conduits are first laid in position and the conductors or wires drawn through

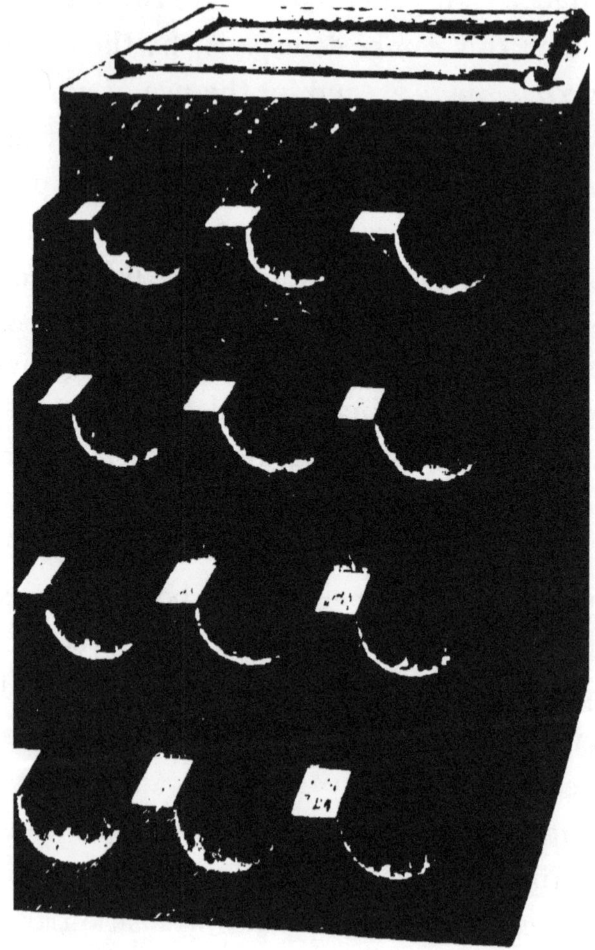

Fig. 35. Wooden conduits for electric conductors. Twelve ducts.

them from man-holes provided for the purpose. For this reason the conduit, as far

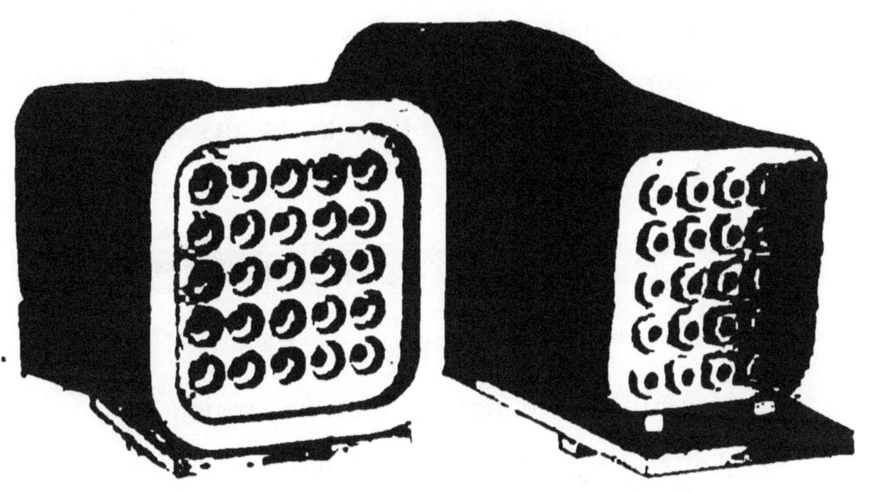

Fig. 36. Glazed stone-ware conduit.

as possible, must extend in a straight line from man-hole to man-hole.

The system of distribution of electric mains by electric tubes is in very common use by the Edison Electric Illuminating

Companies in the principal cities of the United States. In the *three-wire system of distribution*, employed when the lamps are connected in multiple-series, these electric tubes consist of three conductors placed in an iron pipe and insulated from one another by means of a

Fig. 37. Edison electric tube with coupling box.

bituminous insulating material which is poured into the tube while hot.

An electric tube containing three separately insulated wires, provided with what is called a *coupling box* J, is shown in Fig. 37. The coupling box is provided in order to permit the ready joining of the conductors in the separate lengths of tube. The separate tubes are first laid in position in trenches, dug in the street, and the

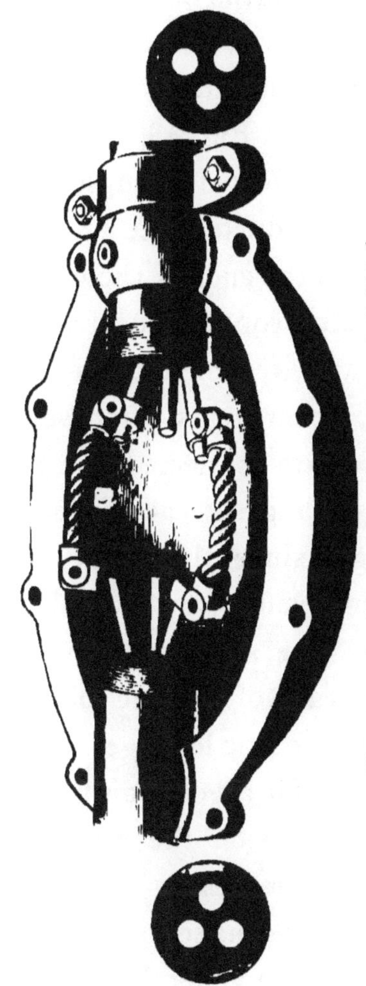

Fig. 38. Coupling-box of Edison tube showing method of jointing.

separate lengths of conductors electrically connected and joined. This is accomplished at the junction boxes in the manner shown in Fig. 38, where the joints are seen to be effected through flexible stranded copper conductors C, C, which are connected to the ends of the wires in the tubes by electric connections and then soldered.

In any system for the distribution of water, steam or gas, where separate lengths of pipes are connected, the joints must be made with great care, in order to prevent leakage, since the water, steam or gas pressure would cause a leak at a defective joint. The same care must be exercised in the case of electric joints, in order to prevent electric leakage. In any case, the necessity for a good joint will increase as the pressure on the pipes, tubes, or conductors increases.

Joints or connections in water, steam or gas pipes are generally obtained by means of screw threads, placed on the ends of the pipes, which are then screwed into each other, the joints being generally completed by means of some form of luting or packing. In Fig. 39, a common form of con-

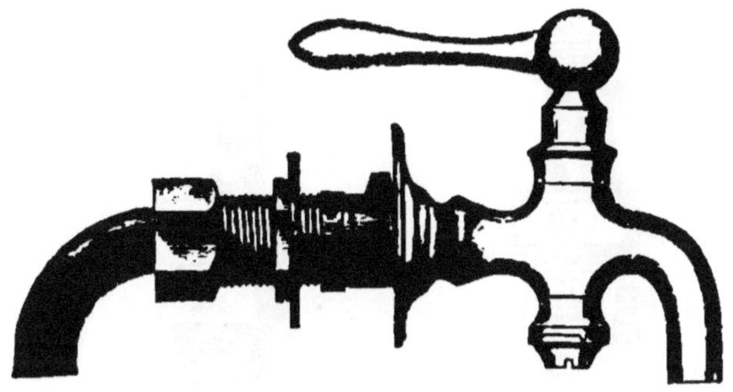

Fig. 39. Connection of water faucet with supply pipe.

nection for a water faucet is shown. Here the joint is effected by means of screw threads. Some forms of joints or couplings for pipes, are shown in Fig. 40, both for straight pipes, and for elbows.

An inspection of Fig. 41, will show the method of connections of pipes at elbows.

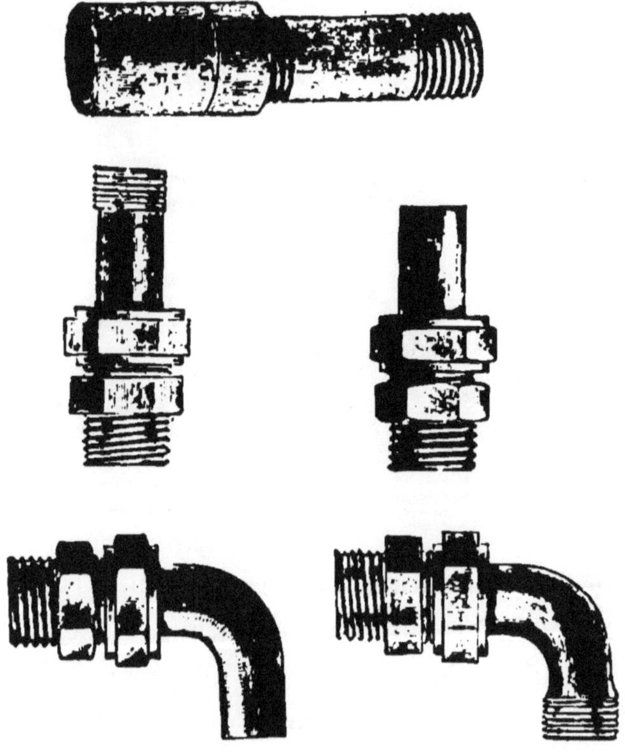

Fig. 40. Forms of joints and couplings for pipes.

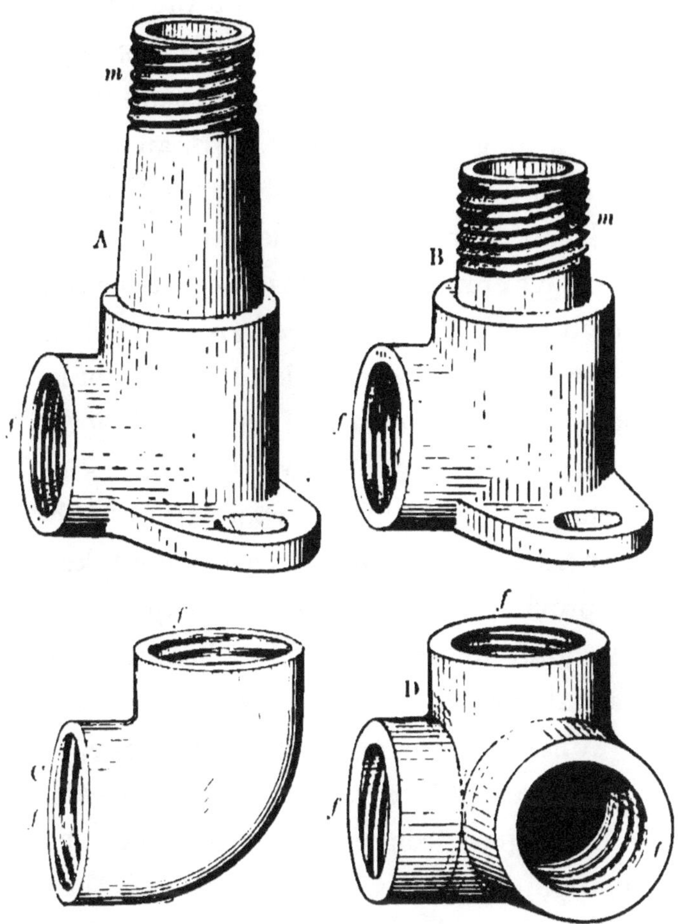

Fig. 41. Form of elbows in pipe.

Screw threads are provided on the outside of one end, as at m, and on the inside of the other end as at f. At A, is an elbow with a long outlet piece; at B, an elbow with an outside and inside screw thread; at D, an elbow with a side outlet.

Under ordinary circumstances it is often much easier to form an *electric joint*, than a steam, gas, or water joint. It is only necessary, in the case of an electric joint or connection, to bring two clean pieces of metal, or other good conducting substances, into firm contact, care being taken to provide sufficient carrying capacity at the joint to prevent undue heating and partial stoppage of the current.

Where the current is not very large, as in the case of telegraphic circuits, a sufficiently good connection is obtained by wrapping the ends of the wires together, as shown in Fig. 42, which represents a common form of *telegraph joint*. A common form of electric connection called a *binding post* is shown in Fig. 43. Here the ends of the conductors are inserted in openings at o, o, o, etc., and afterwards securely clamped in position by means of the screws S, S, S, etc. Similar forms of

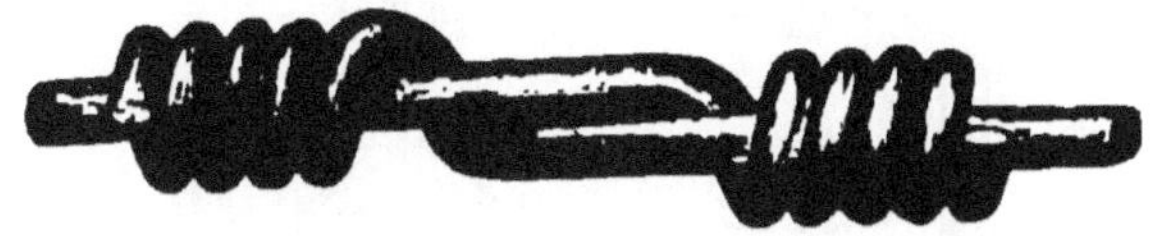

Fig. 42. Telegraph joint.

electric couplings or *connectors*, suitable for larger wires and a stronger current, are shown in Fig. 44.

The fact that an electric connection can be made by merely bringing two conducting pieces of metals into firm contact, renders it very easy to make such connections. Merely bringing the ends of two conductors together, and holding them firmly in contact will suffice. In the well known case of an *electric push*

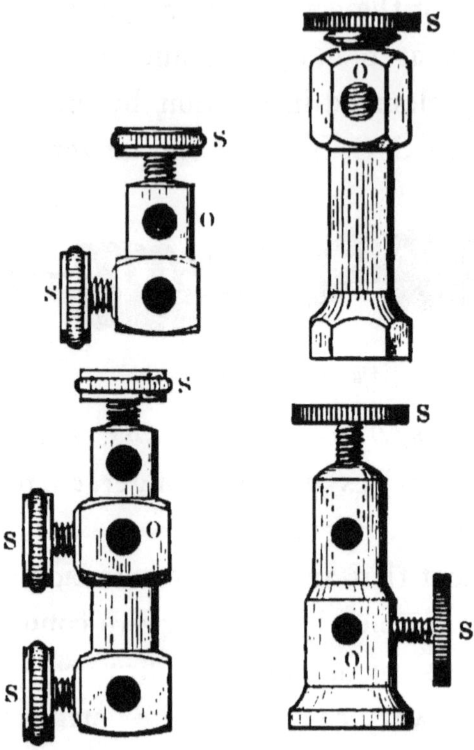

Fig. 43. Forms of binding posts for electrical connection

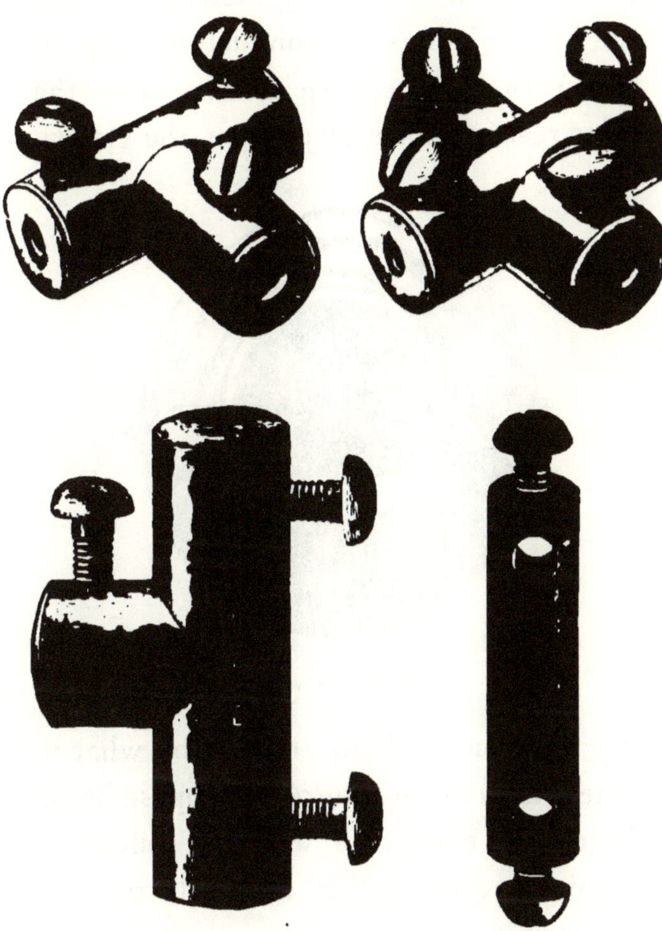

Fig. 44. Form of electric couplings.

button shown in Fig. 45, merely pressing the button brings two conductors together and turns the electricity on. For the same reason, it is easy to turn the electri-

Fig. 45. Push button.

city from one path to another by what is called an *electric switch.* Suppose it is desired to connect a wire or conductor terminating at S, Fig. 46, with conductors terminating at 1, 2 or 3. It is only necessary to join S, with a metallic lever

L, pivoted at S, and provided with an insulating handle. When L, is moved so as to touch 1, 2 or 3, it will join or electrically connect S, with 1, 2 or 3, respectively.

Fig. 46. Electric switch.

While it is easy, as we have seen, to make an electric joint by merely bringing two bare and cleaned wire ends into contact, yet a permanent joint should always be soldered so as to establish permanent metallic connection and prevent the influence of dust or oxide from interfering with the connection, and as already mentioned, such permanent joints should be carefully covered with insulating material, especially when connected with high-pressure systems.

CHAPTER IV.

HOW THE STREET MAINS ARE SUPPLIED WITH ELECTRICITY.

In tracing the water pipes from the house we found that they terminated in a reservoir kept filled with water by the action of a pump; so too, we found that the gas pipes terminated in a gas reservoir, and that the system of hot water pipes terminated in a boiler. In all of these cases the water, gas, steam or hot water so stored was driven or caused to flow into the pipes by reason of water, gas and steam pressure. We would naturally expect then, by analogy, to find something in the central electric station corresponding electrically to the water reservoir, to the gas-holder, or to the steam and hot water boiler. Nor will we be disappointed in this expectation. Tracing the wires as they enter the station from the street, we will find that they pass to what is called a *switch-board* by means of which they are connected with what are called *dynamo-electric machines*, *electric generators*, or *dynamos*. The dynamos produce the electricity which supplies the conductors in the street, the service wires, and the various risers and conductors in the houses.

Taking the analogy of the water supply, the dynamo corresponds to the pump which keeps the reservoir filled with water. The dynamo, may, therefore, be regarded as an *electric source*. Strictly speaking no dynamo or other electric source produces electricity. What it really produces, is *electric pressure*, or a variety of force that is generally called *electromotive force*. It is this electromotive force or pressure that causes electricity to flow through the conductors connected with the source, just as a water pump produces a water pressure, or *water-motive force*, which causes the water to flow through the water pipes. As the term electromotive force is frequently used in books on electricity, it is found convenient to contract it to E. M. F.

Many separate mains enter the central station from the various streets supplied by the station. These are all connected with very large pairs of copper bars called the *bus-bars*. As the electricity flows out of the bus-bars into the mains, and into the conductors connected with them, the electric pressure would tend to decrease or fall, just as the water pressure would tend to decrease or fall in the water pipes, as the water runs out of the reservoir. The electric pressure, however, is maintained by the action of the dynamos.

We will now inquire how the dynamos are driven or moved. There are generally two ways in which central-station dynamos are driven; viz., by *water-power*, and by *steam-power*. Where a good, reliable water-power can be had, water-power is frequently employed. Fig. 47, shows a form of *water-wheel* called a *turbine*, that is frequently employed in central stations. The water which drives the wheel is delivered through the iron flume or penstock P. There are two turbines here employed; one at T, and one at T′. The driving shaft S, has a pulley W, attached to it in the position shown. In large cities, steam power is usually employed to drive the dynamos. In such cases the dynamo shaft is generally coupled directly to the engine shaft; and the dynamo is called a *direct-coupled dynamo*. A pair of direct-coupled dynamos, suitable for central station work, are shown in Fig. 48. The

Fig. 47. Pair of sixteen-inch turbines for water power.

Fig. 48. A direct (General Electric) coupled dynamo, suitable for central-station work.

dynamos are shown at D, D′, directly connected to the engine shaft. These dynamos are capable of feeding a great number of electric lamps and are, therefore, suitable for central-station work. Where fewer lamps are to be used, a smaller dynamo is employed, as for example, that shown in Fig. 49, which is directly coupled to a steam engine. Here the dynamo has six projections around which coils of insulated wire M, M, M, are wrapped, forming what are called six magnetic poles. Metallic brushes B, rest on the commutator and carry off the current. Conductors C, C, lead the current from the machine to the switchboard, and thence to the lamps and other devices fed by it.

There are many varieties of electric sources; but, for central-station work, the dynamo is universally employed. The dynamo is made in a great variety of forms, but, generally speaking, the part that revolves is called the *armature*. This is the part marked A, in Fig. 49, and magnets M, M, M, etc., are provided with masses of iron called *pole-pieces*, curved so as to conform to the outline of the arma-

Fig. 49. Dynamo directly coupled to engine and suitable for small electric station, or for an isolated electric plant.

ture. These magnets are called the *field magnets* and consist of cores of soft iron wrapped with coils of wire. In order to make the electricity pass through the entire length of the wire in the coils, the wire is wrapped with cotton, or some other substance which prevents the electricity from passing out of the wires at its sides; in other words, the wire is covered with an insulating substance. The armature consists of a *laminated core*; i. e., a central part called a *core* formed of plates of soft iron, wrapped around with insulated wires. A device called the *commutator* enables the currents generated in the armature, to be supplied through the brushes in the external circuit in the same direction.

We have now traced the electricity back from the lamps in the house through one of the mains into the street to the dynamo at the central station, and back again through the other main to the lamps. There is thus provided a conducting path between the dynamo and the lamps. This conducting path is called a *circuit*. Although the word circuit means primarily a circular path, circuits are rarely circular in their course, and only partake of the circular character in so far as the circuit forms a path that returns to the first place from which it started ; viz., at the electric source. For convenience, the electricity is assumed to leave the dynamo and flow through one of the mains to the lamps placed in the circuit, and then to flow back again through the other main to the dynamo. The *pole* or *terminal* of a dynamo, or other source out from which the electricity is assumed to flow, is called the *positive pole* or *positive terminal*, and that at which it returns to the dynamo or source, after having passed through the circuit, the *negative pole* or *negative terminal.*

In Chapter 2, we have referred to the fact that the carrying capacity of a water pipe, which of course will depend upon its size, must increase as the quantity of water which flows through it increases. We will now examine more in detail, how the quantity of electricity which a given wire or conductor can safely carry, can be determined. Substances vary greatly in the readiness with which they permit electricity to flow through them; or, in other words, they differ in what is called their *conducting power.* Metals offer but little resistance to the passage of electricity, and are called *conductors*, while other substances, like air, glass and hard rubber, offer an enormously great resistance to its passage and are called *non-conductors* or *insulators.* We say that the *electric resistance* of metals is low, and the electric resistance of air, glass, or hard rubber is high. The electric resistance of a substance is measured in units called *ohms*, from Dr. Ohm, a German physicist. An ohm is the resistance of about 2 miles of ordinary trolley wire; or of about one foot of very fine copper wire of No. 40, A. W. G. (American Wire Gauge).

In wires or conductors of the same kind of material, the electric resistance increases with the length of the wire. Thus, 2 feet of wire have just half the resistance of 4 feet of the same wire. The electric resistance also increases with a decrease in the diameter of the wire. For example, if in Fig. 50, a certain length of wire, say that at A, has a resistance of one ohm, then, if this wire be cut in half, each half

Fig. 50. Effect of length and cross-section of conductors on their electric resistance.

as at B, would have a resistance one half as great, or half an ohm; and if these halves be placed together, or connected in multiple, as shown at C, their joint area of cross section would be doubled, and their joint resistance reduced one half; i. e., to one quarter of an ohm.

When a water pressure is permitted to

act upon the water in an open pipe, it will cause a flow or current of water through the pipe, which will be greater, the smaller the resistance of the pipe; that is the shorter the length of pipe and larger the area of cross section of pipe. If the pipe be long and narrow, only a small current of water will flow through it, since the resistance it offers to the flow is great.

When an electric pressure or E. M. F. is permitted to act on a complete circuit, it will cause a *flow of electricity*, or an *electric current* to pass through that circuit. The smaller the resistance of the circuit, that is the shorter its length, and the larger the wire of which it is composed, the greater will be the current which a given electric pressure can produce. A circuit formed by a long, thin wire, will only permit of a feeble current passing through it, owing to its high resistance. When the resistance of a circuit is high, the electric pressure must be great in order to cause a strong current to pass through it, while if the resistance of the circuit be small, a much smaller pressure will be required.

We measure electric current in *units of current* called *amperes.* One ampere will flow through a circuit whose resistance is one ohm, under an electric pressure of one volt. If the resistance of a circuit be known in ohms, and the E. M. F. or pressure acting in that circuit be given in volts, the current in amperes may be obtained by dividing the volts by the ohms.

CHAPTER V.

THE ELECTRIC LIGHTING STATION.

LET us enter the lighting station and examine some of the machinery it contains. Probably the first thought that suggests itself, is that a great amount of power has to be employed, to drive the dynamos or electric generators. We will suppose, as is usually the case in central stations of large cities, that steam power is employed. In this case, the three general types of machinery the station contains, are the boilers, the steam engines, and the dynamos or generators. If the time of our visit is during the busy hours in the evening, when the station has its greatest load, we will find the fires burning brightly under the boilers, which are generating plenty of high-pressure steam, as a glance at their steam gauges will show. There can be no doubt that this steam is doing work; for, the engines are running at full speed driving the dynamos and furnishing electricity.

All this expenditure of energy is necessary in order to produce the electric pressure and cause the electricity to flow from the station through the street mains; to enter the houses at the service wires; and to pass through the risers into the lamps, wherever they are turned on; for, whenever the electricity flows through a conductor it is accompanied by electric energy, which it expends in the circuit in producing light or heat, or some other kind of electric effect. In the case we are considering, the energy is derived from the burning coal. This energy is liberated as heat energy, when the coal is burned, and is transferred in the boiler from the burning coal to the steam. The energy in the steam, is, in turn, transferred to the engine in which it produces energy of motion. The engine transfers its energy of motion to the dynamo, which produces electric energy, and this energy passes through the street mains or conductors, into the houses, and finally appears as luminous energy in the light of the incandescent lamps when they are turned on.

In all of the above cases there are transformations or conversions of energy. We start with a certain definite amount of energy in the coal, each ton of coal of a given quality containing a definite amount of energy, capable of being measured in units of energy. Under no possible circumstance can the amount of energy which is transferred to the boiler, during the burning of one ton of coal, exceed the energy in that ton of coal. On the contrary, there is always considerable energy lost or wasted; as for example, by the heat which escapes up the chimney, so that the energy transferred to the boiler is always much less than that contained in the coal burned; then there is another loss in transferring the energy from the boiler to the engines; still another loss of energy occurs in transferring the energy of the moving steam engine to the dynamos, so that when the electrical energy is finally generated, by the burning of say one ton of coal, the amount of electric work the current is capable of performing is very much less than the energy that is known to be originally contained in that ton of coal.

But the loss of energy does not stop here. When the electric pressure causes electricity to flow through the mains in the streets, and through the risers or other conductors in the house, some energy is again lost; and, finally, when the electricity passes through the incandescent lamps, another and very considerable loss occurs, the energy contained in the light which appears being very much less than that present in the current which causes it, nearly all the energy liberated in the lamp filament being expended in producing heat instead of light.

It is necessary, therefore, in the management of a central station, in order to ensure profitable working, to employ machinery that is economical in operation; that is, in which the losses are as small as possible; or, as it is usually called, to employ machinery possessing as high an *efficiency* as possible, since by this means the amount of coal consumed for a given amount of electric lighting can be materially reduced.

As the word efficiency is very important, we will explain its meaning somewhat at length. In the case of a steam engine, the amount of energy required to drive the engine is called the *intake* of the engine. The amount of useful energy the engine gives out, that is the amount of energy that is capable of being applied to some useful purpose, say for driving dynamos, is called the *output* of the engine. The *efficiency* of the engine is equal to its output divided by its intake. Were it possible for any machine to be produced in which no losses were experienced in driving it, the output would be equal to the intake. Consequently, the output divided by the intake would be 1, or 100 per cent. In all cases, however, the output being smaller than the intake, the efficiency is less than 100 per cent. If, for example, one half the energy were lost in any case, so that the intake was 100, and the output 50, then,

$$\text{The efficiency} = \frac{\text{Output}}{\text{Intake}} = \frac{50}{100} = .5 \text{ or}$$

50 per cent. The efficiency of a good dynamo is very high, being in large sizes equal to 95 per cent. or even more. The efficiency of engines is much lower. For this reason large central stations generally employ what are called *double* or *triple-expansion engines*, since the engine efficiency is thereby increased.

If we could enter the central station at the place where the wires or conductors pass into the street, we would find, as already mentioned, all these wires ending in large conductors called the bus-bars. The word bus, is a contraction for *omnibus* meaning "*for all*," this word being given to these conductors because they carry all the current generated in the station by the dynamos.

Suppose, for example, the station entered was that shown in Fig. 51. Here, we should find two generators directly coupled to the driving engines. In a smaller station we might find the dynamos *belt-driven*; i. e., connected to the driving engines by means of belting.

Fig. 51. Small central station.

The loss in transmission in belt-driving is somewhat greater than in direct driving, so that in large central stations direct driving is generally employed.

The two positive terminals of the dynamos are connected to the *positive bus-bar*, and the two negative terminals to the *negative bus-bar*; the bus-bars, as we have seen, being connected to the feeders, and through them to the conductors connected with the mains. Between these, and governing the flow of electricity from the central station, is a very important device called the *switch-board*. This is shown on the right hand side in the figure.

Although the switch-board of a large central station is a complicated piece of apparatus, yet its general operation and construction are not difficult to understand. As is well known, in the case of the gas supplied to a house, a meter is necessary to measure the quantity of gas that passes into the house from the street mains. So, also, in the case of the gas supply, a large meter is inserted between the gas-holder and the mains; and, in the case of the boiler that supplied the steam radiators, there is a gauge to indicate the steam pressure. So in the case of the central station, a device called an *ammeter* is necessary to measure the quantity of electricity that passes from the station to the mains, and also a device called a *volt-meter*, to measure the pressure which the dynamos maintain at the bus-bars. These are placed on the switch-board, and are shown in the figure at the upper row R, R. G, is a steam gauge.

The switches required in central stations, where large electric currents are to be transferred from one circuit to another, are formed of heavy pieces of copper, and, are furnished with several contacts. There are generally many of these switches on the switch-board. One these switches is shown in Fig. 52. Here, one side of the circuit is connected to heavy terminals at 1, 1, and the other side to terminals at 2, 2.

The switch is represented in the figure as being closed. In order to open it, the

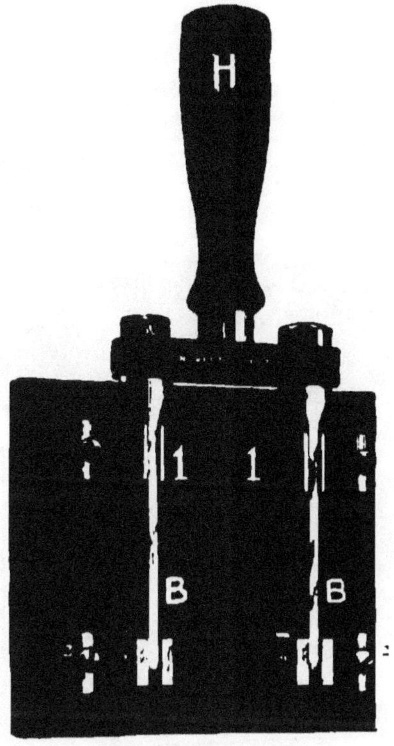

Fig. 52. Double-pole switch.

switch handle H, is moved forward so as to cause the metallic knife blades B, B, to leave the spring clips 1, 1. In order to ensure good contact, the width of the spaces in these clips is somewhat smaller than

Fig. 53. Weston ammeter.

the width of the knife blades B, B, which on entering, push apart the springs.

Figs. 53 and 54, show forms of ammeters and voltmeters suitable for central station

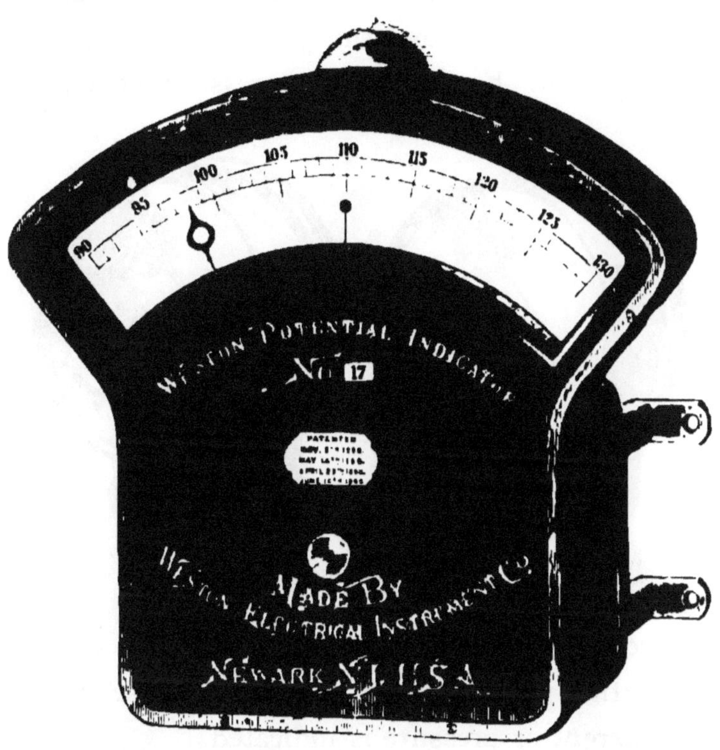

Fig. 53. Weston voltmeter.

work. In the form shown in Fig. 53, the end of the needle only is shown on the graduated scale, in a position indicating that no current is flowing through the circuit, with which it is connected. The pointer on the voltmeter indicates an E. M. F. or pressure of 98

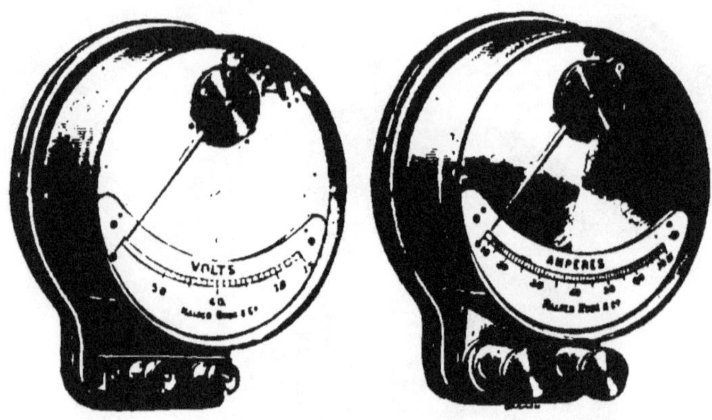

Fig. 54. Nalder voltmeter and ammeter.

volts. In the form shown in Fig. 54, the entire pointer is seen. In this case no current or pressure is indicated.

A switch-board suitable for a smaller station is shown in Fig. 55. Here the ammeter, voltmeter and switches are more

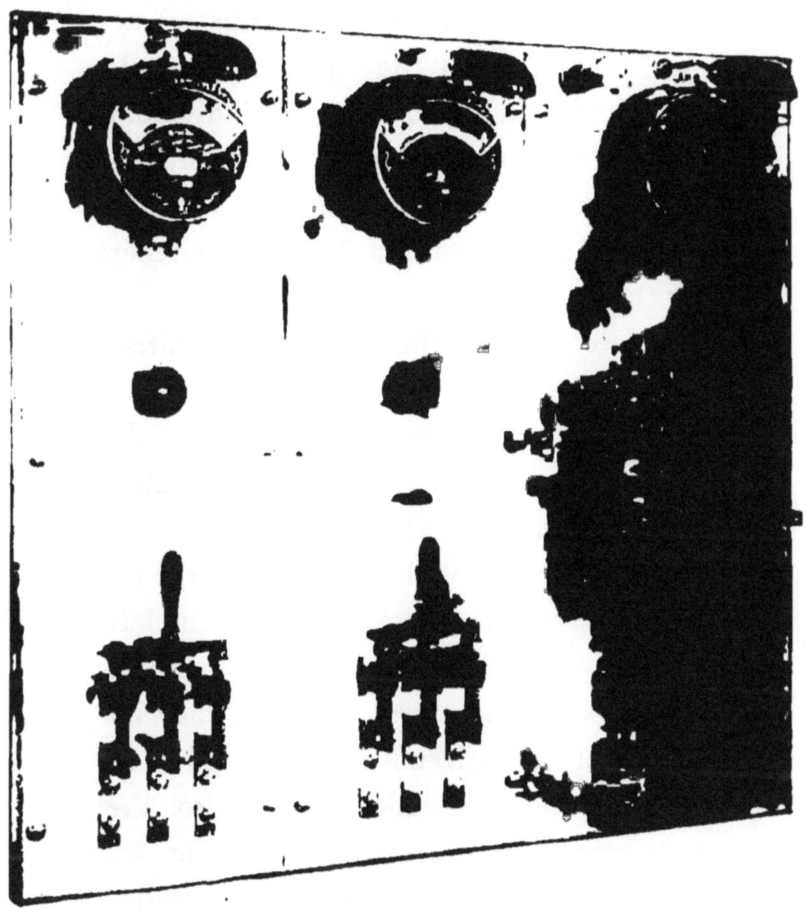

Fig. 55. Switch-board for isolated lighting plant.

clearly seen. This switch-board consists of three *panels* or vertical strips. The small switches, on the right-hand panel, are for closing and opening individual circuits, each connected with a comparatively small number of lamps. The two left-hand panels contain the main switches for the dynamos.

Before closing the chapter on central stations it may be well to examine briefly the stations shown in Figs. 56 and 57. In Fig. 56, we see a part of the Edison Electric Illuminating Company's station at Brooklyn, N. Y., with two pairs of large direct-connected dynamos. Those on the left are 8-pole generators, and those on the right are 14-pole generators. Fig. 57, shows a station of the Citizens' Electric Light Co., at Leadville, Colorado. Here, also, the generators are direct-driven. In Fig. 57, the switch-board is shown on the right-hand side of the figure. For the reason

Fig. 56. General Electric Company's generators in the Edison Electric Illuminating Company's station, Brooklyn, N. Y.

Fig. 57. General Electric Company's ironclad generators, Citizens' Electric Light Co., Leadville Col.

Fig. 58. Walker Company's incandescent dynamo, suitable, for isolated plants.

already given, belt-driven generators are seldom employed in large central stations. For isolated plants, however, and in cases where steam power is already installed, belt-driven generators are employed. Fig. 58, shows a belt-driven generator of the Walker Company, suitable for an isolated station.

CHAPTER VI.

HOW THE INCANDESCENT LAMP OPERATES.

THE incandescent electric lamp belongs to that class of devices in which the energy of the electric current is transformed into heat. Considered merely as a heating device its efficiency is perfect. But we use the electric lamp not for the heat it can furnish, but for that portion of the heat which produces light. It is a well known fact that when a body is sufficiently heated it will begin to glow or emit light as well as heat. When the temperature of about 932°, on the Fahrenheit scale, is reached, a heated body will give off red light; or, in other words, becomes *red hot.* As the temperature is increased still further, the hot body throws off successively orange, yellow, blue, green and violet light. At about the temperature of 2730° F, it emits all the colors of ordinary sunlight, or is *white hot.*

But when a heated body glows or emits light, it also emits heat, and, unfortunately, the amount of energy absorbed to produce the heat is much greater than the amount which produces the light, so that the electric lamp has an exceedingly low efficiency as a producer of light. In, perhaps, the best incandescent lamps known, only about four per cent. of the total energy in the electric current produces light, the remaining ninety-six per cent. being uselessly expended in producing heat. This is true of all artificial illuminants; they produce far more heat than light.

The higher the temperature to which a solid is heated, the greater will be the proportion of light produced to heat produced. Consequently, the higher the temperature to which the filament of an incandescent lamp can be raised, the greater will be the amount of light it will produce with a given amount of heat; or, in other words, the greater will be its efficiency as a producer of light. There is no difficulty in raising a carbon filament to a high temperature. To do this it is only necessary to send a sufficiently strong electric current through it, and this is done by increasing the pressure or voltage at the terminals, just as we can increase the quantity of water, steam or gas passing through a pipe by increasing the pressure. Indeed, we can readily increase the temperature of the filament by this means to such an extent as to volatilize it, but this would of course ruin the lamp. Consequently, there is, in every lamp, a point beyond which its light-giving power cannot be safely passed. If the lamp is burned at a lower temperature than this point, the light it emits is small in quantity, and is of a reddish tint or color. If the lamp is burned at a higher temperature, the quantity of light it emits is greatly increased, and the light approaches more closely to the character of sunlight, but the life of the lamp will be greatly decreased.

Now the circumstance which determines the temperature of the lamp filament, and, consequently, the amount and character of the light it emits, is the quantity of electricity which passes through the filament; or, indirectly, the pressure or voltage at the lamp terminals. With a given diameter and length of carbon filament, there will be a fixed voltage at which it is necessary the lamp shall be operated, in order to obtain its greatest efficiency with a given length of life. In most cases this voltage is marked on the lamp.

In the United States, incandescent lamps are usually operated either at about 50, or at about 110 volts. By properly proportioning the dimensions of the filaments, lamps can be operated either at much lower or at much higher pressures. The incandescent lamp shown in Fig. 59, is a 16-candle-power, 110-volt lamp; i. e., when supplied at its terminals with a pressure of 110 volts, it will give an amount of light equal to 16 *standard candles*, each burning at the rate of 120 grains per hour. Nearly all incandescent-lamp circuits operate either at a pressure of 110 volts, or of 50 volts. Though lamps could be made to operate on higher voltages than 120 volts, this is rarely done. Lamps of lower voltages than 50, however, are common. Fig. 60, shows a 16-candle-power, 50-volt lamp, Fig. 61, shows a number of *miniature electric lamps*. These are made for various pressures,

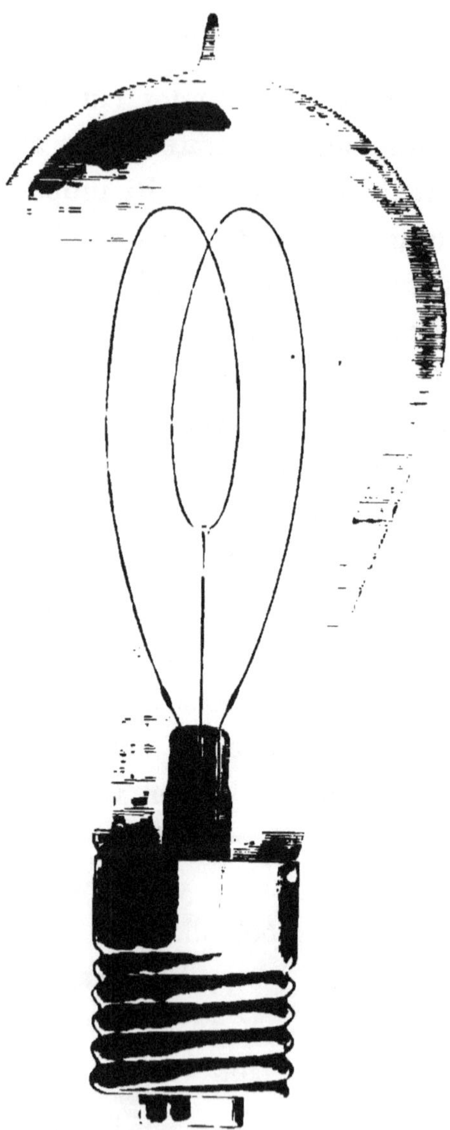

Fig. 59. 16-candle-power, 110-volt lamp.

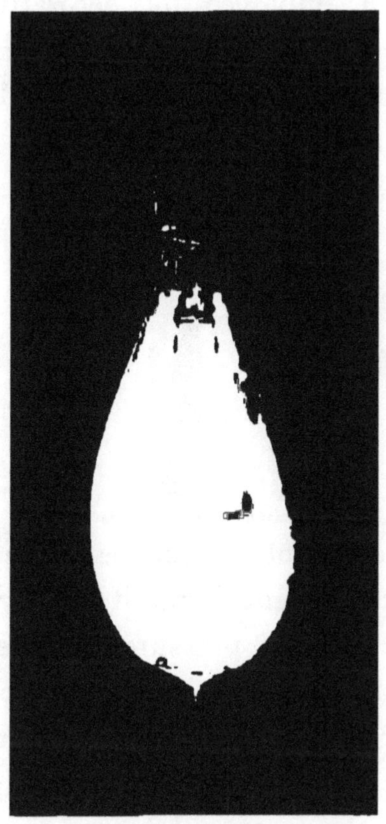

Fig. 60. 16 C. P. 50 Volt Sawyer-Man lamp.

usually 25, 30, or 50 volts. Such lamps are often employed for decorative purposes. Still smaller lamps are made for lower pressures down to a single volt. These are suitable either for electric jewelry, for microscopes, or for bicycles.

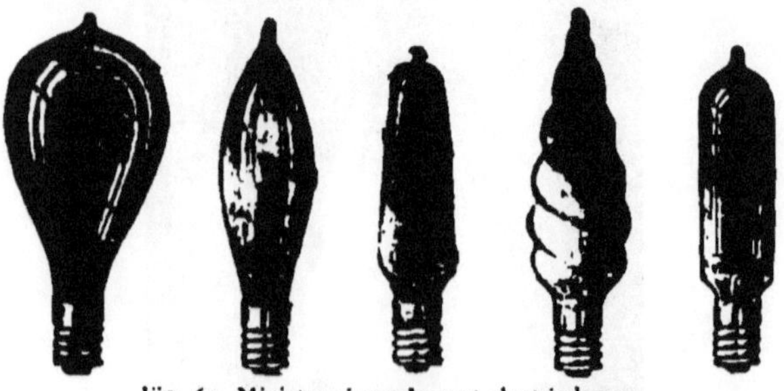

Fig. 61. Miniature incandescent electric lamps.

In order to obtain the best service from any incandescent lamp, it is very necessary that the pressure at the lamp terminals be maintained at the pressure for which the lamp was manufactured. If the pressure is too low, the current passing through the lamp will be too small, and, consequently, the temperature of the filament will be too low, so that the light emitted will be small in quantity and of a reddish color. The lamp, however, will last under these conditions a very long time. On the other hand, if the pressure at the lamp terminals be too high, the current passing through the lamp will be excessive, the temperature high, the

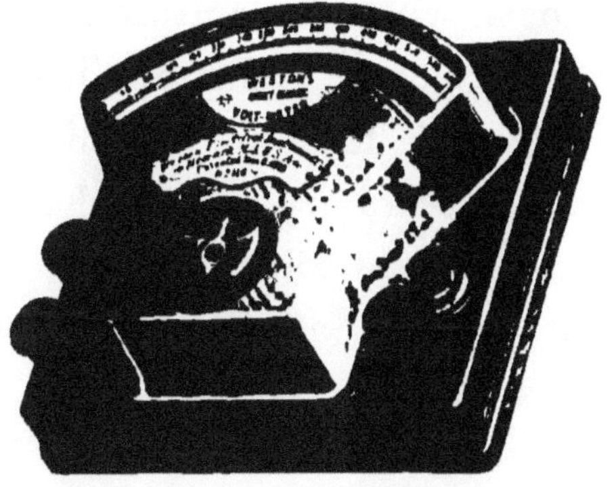

Fig. 62. Weston standard portable voltmeter.

quantity of light emitted great, and the color more nearly approaching that of sunlight, but the *life of the lamp* will be greatly diminished; i. e., the lamp filament will soon break. Too high a voltage at the lamp thus means a short but brilliant life. Too low a voltage, a prolonged but dismal life. In order to test the voltage a form of portable voltmeter is used, such for example, as that shown in Fig. 62.

It is not entirely a matter of the length of life of the incandescent lamp, at what voltage or at what temperature it is operated. The colors of bodies are due to the light by which they are illumined. A red body appears red, only when it is lighted with light containing that color. Sunlight contains all the colors of the rainbow, as may be seen by looking at a patch of sunlight through a prism. When sunlight falls on a red leaf, or on a piece of red cloth, all the colors except the reds are absorbed, and the leaf or cloth throws off or emits the reds only. If we illumine the leaf or cloth with light which contains no reds, it will appear nearly black. This we can do by burning a piece of lamp wick, soaked in alcohol in which some common table salt has been dissolved. The light given off by the burning wick consists of a nearly pure yellow, so that red objects examined by it in a darkened room appear black. In a similar manner, yellow, green or blue objects can appear at night in their true tints, or *daylight colors*, only when examined by light that contains their particular yellows, greens and blues. If then, we wish the light of an incandescent lamp to give to colored objects their true sunlight colors at night, the light must approach as nearly as possible the characteristic color of sunlight, and this is more closely approached, the higher the temperature at which the lamp is operated.

But even if we operate the lamp at the actual pressure for which it was made, it will be found to gradually fail, so that, without breaking, the lamp may become worthless. During the first few hours of operation the lamp actually increases in brightness; but this soon ceases, and the lamp begins to deteriorate, and, at the end of, say 2,000 hours, may become practically useless. There are several causes of this deterioration. During use, the high temperature at which the lamp has been operated has caused the carbon to gradually disintegrate, minute particles being thrown off and deposited on the walls of the lamp chamber. In other words, the carbon filament gradually evaporates. A double injury is thus produced; the carbon filament is reduced in size, so that the pressure for which the lamp was calculated will be no longer sufficient to properly operate it; moreover, the carbon deposited on the walls of the lamp chamber produces a blackening which obscures the lamp by preventing the light from passing out. When a lamp reaches this stage it should be replaced by a new one.

In order to obtain the best illumination of a room, or other apartment, it is neces-

Fig. 63. Combination gas and electric bracket.

sary that the light be well distributed. Where all the light is concentrated in a single point, bad shadows are produced. To avoid these it is necessary to employ a

Fig. 64. McCreary's silvered reflector shade.

number of separate lamps, so placed that each lamp throws light on the shadows produced by the other lamps. The same object can be attained, for a limited area, where a single lamp is used, by covering the lamp by a shade G, as is shown in Fig. 63, in the case of a *combination gas and electric fixture;* or, the lamp may be surrounded by a *silvered reflector shade* S, as shown in Fig. 64. Or a *full shade* S, or a *half shade* S', may be employed as shown in Figs. 65 and 66. In

Fig. 65. Full shade for lamp.

Fig. 66. Half-shade or reflector for lamp.

all these cases the object is to increase the surface from which the light is ultimately thrown off. Sometimes, where it is desired that the lamp should throw its light

Fig. 67. Half reflector shade.

in one direction only, a *half reflector shade* H, is employed, as shown in Fig. 67.

A form of reflector R, suitable for a desk lamp, is shown in Fig. 68. Here

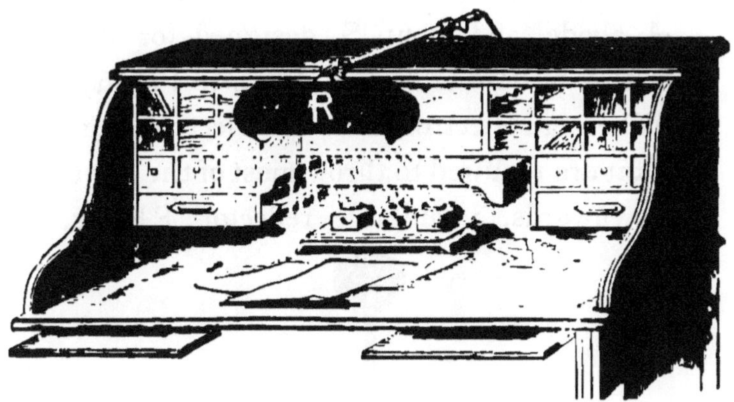

Fig. 68. Reflector for desk lamp.

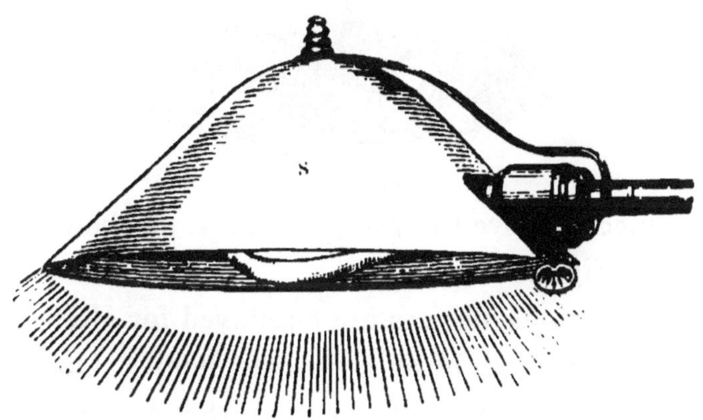

Fig. 69. Shade and lamp for billiard table.

the light is thrown directly on the desk. A shade, and lamp S, designed for use with a billiard table, are shown in Fig. 69. When an electrolier, or an electric pendant are required to throw most of the light downwards, a suitable reflector RR, is

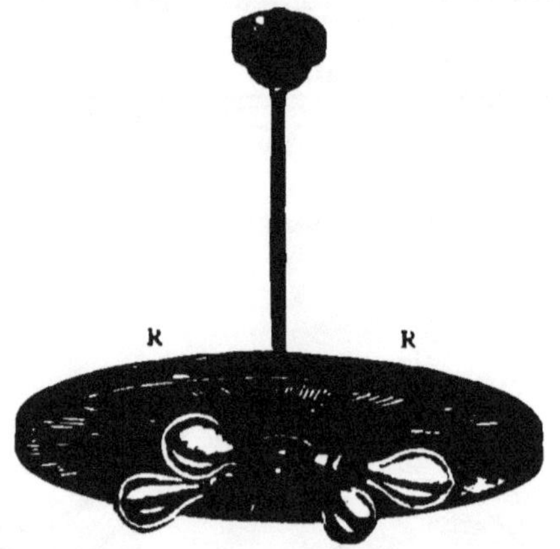

Fig. 70. Frink's electric cluster reflector.

placed above the group of lamps as shown in Fig. 70.

Various devices are employed for altering the height of a pendant lamp. One of the simplest of these is the *ball-cord adjuster*, where by merely increasing the length of the loop L, Fig. 71, by pulling more of the flexible *twin-cord conductor* into the ball, raises the lamp, while decreasing the length of the loop, lowers the lamp.

Fig. 71. Ball cord adjuster.

Where lamps are in exposed places, and are, therefore, apt to receive blows, they are sometimes covered by a *wire guard* in order to protect them. Of course, these guards throw shadows, but this is not very objectionable in most places where such lamps are used. A full and half-wire guard are shown in Fig. 72, and a lamp

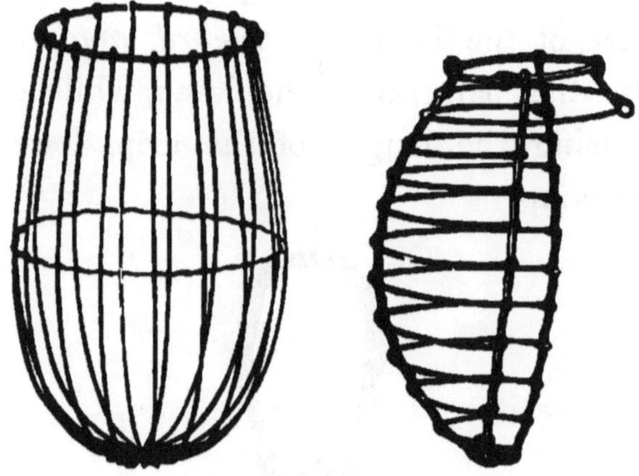

Fig. 72. Lamp wire guards.

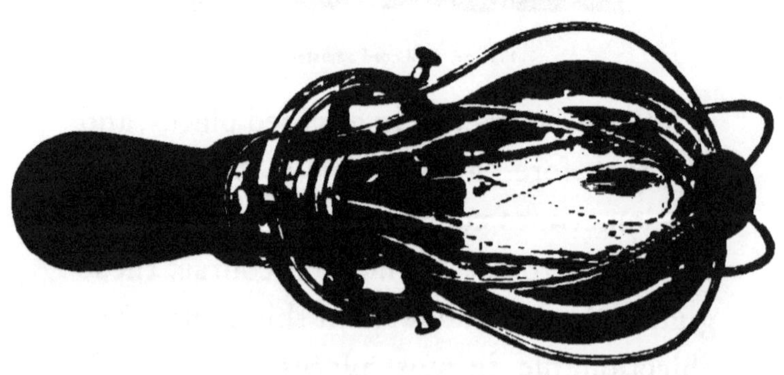

Fig. 73. Lamp guard with lamp in place.

provided with such a guard in Fig 73. The lamp shown in Fig. 73, is provided with a handle ; it is what is called a *portable hand lamp* ; i. e., a lamp which can be

Fig. 74. Portable lamp stand.

carried to such a short distance as the length of its supply wires may permit.

A form of portable electric lamp with stand, is shown in Fig. 74. The flexible twin conductor, which carries the current to the lamp, permits of its motion through a limited distance.

It is sometimes necessary to protect a lamp from the effects of steam or vapor. For this purpose the lamp is placed inside a glass chamber or *vapor-globe* B, Fig. 75. This chamber is sometimes provided with a valve, in order to prevent the expansion of the contained air from breaking the chamber.

As a rule, houses are now built with the electric conductors concealed in the plaster or placed inside iron conduits or tubes. Sometimes, however, it becomes necessary to install an additional circuit. In such cases the conductors are placed inside wooden *mouldings*, such as shown in Fig. 76; or, are supported in wooden *cleats*, of the form shown in Fig. 77.

But the wires or conductors which carry the current to the lamps would themselves become intensely heated, if a sufficiently strong current be passed through them, which would not only instantly destroy all

Fig. 75. Vapor-globe for incandescent lamp.

the lamps in the circuit, but might also set fire to the house. In what manner may this be avoided? Fortunately, the

Fig. 76. Moulding for incandescent lighting circuits.

device for this purpose is both simple and effective. It consists in what is called a *safety fuse* or *safety cut-out.* Its operation

Fig. 77. Cleats for incandescent lighting circuit.

depends on the fact that if a piece of wire, formed of a readily fused lead alloy be placed in the circuit to be protected, should any abnormally strong current pass through that circuit, it would immediately fuse the safety wire or strip, and thus break or open the circuit and protect anything placed in it from the violence of the current. The safety fuse is supported in a block of porcelain or other substance that is not readily affected by a high temperature and is usually covered by

Fig. 78. Porcelain block for safety fuse or cut-out.

another block of porcelain or a sheet of mica. This is necessary; since, when a fuse melts or *blows*, the molten lead is frequently thrown a considerable distance, and fires have been caused in this manner by uncovered fuse wires. A form of porcelain *fuse block* is shown in Fig. 78. The fuse wire is held in position between the screws a and b. Another form of safety fuse, shown in Fig. 79, with the fuse wire inserted between a and b. Sometimes a fuse wire is placed in what is called a *safety fuse plug* as shown in Fig. 80. The size of the safety fuse will, of course, depend on the strength of the current it is intended to carry. Where the current is large, what are called *fuse links* are employed, as shown at A and B, in Fig. 81.

Fig. 79. Safety fuse.

When the pressure or voltage is not kept steady at the lamps, they give an unsteady light or flicker. Such an unsteadiness of pressure may be caused by

Fig. 80. Edison fuse plugs.

an unsteady running of the dynamo at the central station; as, for example, by a loose belt, where the dynamo is belt-driven; or, the unsteadiness may be caused by sudden variations in the currents supplied through the mains. Just

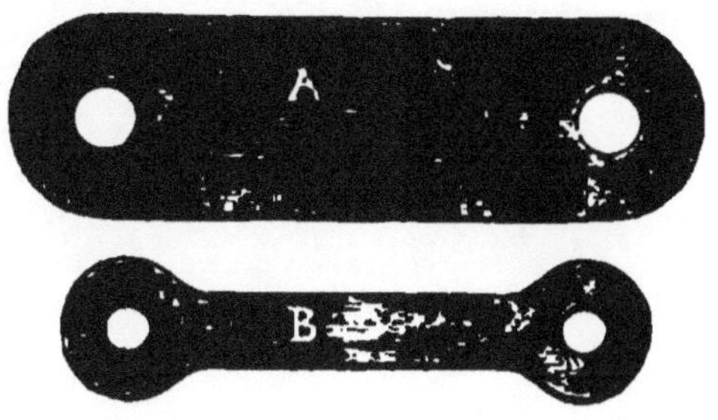

Fig. 81. Fuse links and copper fuse-wire terminals.

as the sudden variations of flow in water pipes will produce fluctuations of pressure in the pipes or water mains; so, sudden variations of current in electric supply mains, will produce fluctuations of voltage in the mains. Thus, when a big electric motor, operated from the electric lighting mains, is suddenly turned off or on, there is apt to be a corresponding slight rise or fall of pressure at the lamps in the neighborhood. The remedy in such cases is usually to increase the size of the electric mains so as to allow them to carry readily a stronger current.

CHAPTER VII.

HOW THE INCANDESCENT LAMP IS MADE.

LIKE all great inventions, the incandescent lamp, as it is made to-day, was not the product of any one mind or any one time. On the contrary, it combines the best thoughts of many men for many years. Looking at the incandescent lamp as it is sent out from the manufacturer, ready to be placed in its socket, it might appear to be a very simple invention; merely a slender filament or thread of carbon placed inside an exhausted chamber of glass, provided with *leading-in wires*, or conductors, sealed in the walls of the chamber by the fusion of the glass around them. It was known, long before the practical lamp was produced, that an electric current sent through a wire would render it luminous, provided the current was strong enough. It was known that carbon was highly refractory; that carbon, when heated in the air, would be rapidly consumed with the oxygen of the air, and that, therefore, to protect a filament of carbon it would be necessary to place it in a chamber from which all air had been removed. But it was one thing to know all these circumstances, and quite another to be able to put them together in an operative lamp. Consequently, it was many years before the commercial incandescent lamp was perfected.

But let us now inquire how the incandescent lamp is made; for, we shall then better understand the circumstances that stood in the way of producing a good lamp. In the first place, it is evident that the material of which the filament is made must be able to stand a high temperature;

for, the higher the temperature to which the filament can be heated, the greater will be the efficiency of the lamp as a source of light. After many unsuccessful trials, carbon was found to be practically the only substance suitable for this purpose. Not only is carbon highly refractory, but it also possesses a high electric resistance in a small bulk or space. Consequently, whereas a copper filament would have to be so thin as to be almost invisible, and therefore impracticable to handle, a carbon filament is thick enough to readily mount and handle. Passing by the long series of experiments made by many able men, we will describe the manner in which the carbon filaments are now generally made. Incandescent lamp filaments were for a long time made from carbonized filaments of bamboo, but are now formed from what are called *squirted filaments.*

In preparing a bamboo filament, the bamboo was first cut and shaped into the desired form. Only certain portions of the bamboo were employed, both the softer inner part, and the hard outer coating being useless. In order to insure a uniform diameter, the bamboo filament, after being cut or shaped, was passed through

A B

Fig. 82. Bamboo filament.

suitable cutting dies. It should be observed, however, that lamp filaments were always cut straight from the bamboo as shown in Fig. 82, being bent into the desired shape before being carbonized.

The length of the filaments varies according to the character of the lamp and the voltage to be employed. The ends of the filaments at A and B, were made larger than the rest in order the better to permit the connection of the filament to the leading-in wires.

The filament now being suitably shaped, was subjected to the *carbonizing process*; i. e., exposed to the action of a high temperature, while out of contact with the air. To do this the filament was bent into the desired curved form by securing it to the surface of a suitably shaped piece of carbon, and then placed inside a retort or carbonizing box, and closely packed with lamp black or other form of pulverized carbon, and subjected to the prolonged action of a high temperature. In other words, the filaments were baked in an oven from which all air was excluded.

Filaments were also sometimes made from suitably prepared carbonized cotton thread.

Squirted carbon filaments have now completely taken the place of bamboo or treated cotton thread filaments. The process consists in forcing finely divided carbon, moistened with a carbonizable liquid, through a die. The carbon thread so obtained is suitably shaped and carbonized.

The filaments now being obtained, the next step consists in *mounting* them ; i. e., connecting them to the leading-in wires which supply them with the electric current and mounting them on supports ready to be placed inside the lamp chamber. The ends of the filament A, A, Fig. 83, are first cemented to platinum wires P, P, and the ends of these connected to copper wires C, C. This platinum-copper joint is effected by fusing the ends together, by holding them for a moment in the flame of a Bunsen burner. A glass tube T, has a shoulder blown on it at S, and its upper end hermetically sealed by the fusion of the glass around the platinum wires P, P.

Much difficulty was encountered, in the

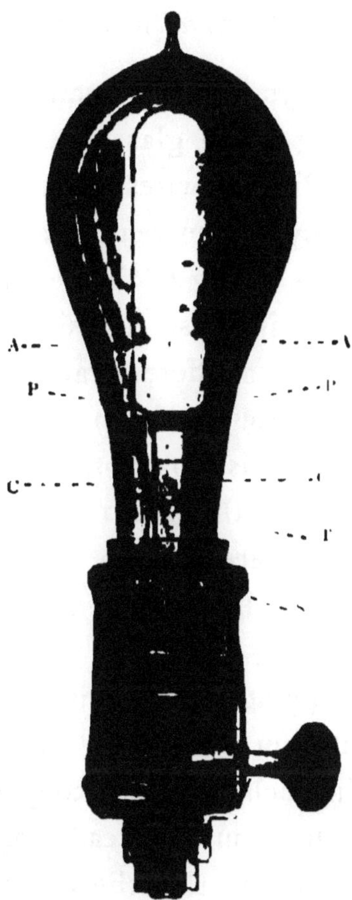

Fig. 83. Mounted filament in lamp bulb.

early days of lamp manufacture, by the breaking of the glass where fused around the leading-in wires when the lamp was heated by the current passing through it. This was owing to the fact that the wire tends to increase in size, as it grows warmer, at a different rate from the glass which surrounds it, so that the seal between them is broken, thus destroying the vacuum in the lamp chamber. It was therefore necessary to use a wire which expands by heat at the same rate as the surrounding glass, and such a wire was found in platinum. Owing to the expense of platinum wire, only short pieces are used in each lamp, the copper wire being joined to them as already explained.

It is evident that the carrying capacity of the leading-in wires must be sufficient to prevent them from acquiring a very high temperature, since this would melt the glass and thus injure the vacuum. The connection of the ends of the filaments to the platinum wires, is a matter of great importance. Various methods have been employed for this purpose. The joints were first made by means of minute platinum bolts and nuts as shown in figure 84; or a minute socket was made in the platinum wire, the ends of the filament imbedded in this socket, and the two firmly secured together by plating the joint with copper, or with carbon. But all these methods have been practically replaced by the simple expedient of joining the ends by a piece of carbon paste, and subsequently carbonizing it by sending an electric current through the joint.

Fig. 84. Filament hermetically sealed in lamp globe.

A difficulty that for a long time stood in the way of the production of a practical incandescent lamp consisted in the fact that when the current was sent through filaments, no matter how great care had been exercised to render them of uniform diameter throughout, they glowed unequally; that is, they were brighter in some spots than in others. If the voltage at the terminals was increased, so as to bring the dull portions to bright incandescence, the temperature of the other portions became too great, and the filaments soon failed. This difficulty was due to the unequal resistance of the filaments, certain portions thus acquiring a higher temperature than the others. It was remedied by a very simple and beautiful invention called the *flashing process*; or, as it is generally called, the *treatment* of the filaments.

The flashing process consists in sending

an electric current through a mounted filament, while surrounded by a carbonaceous gas or liquid. The current first raises the temperature of the parts of the filaments that possess the greatest resistance, to a temperature sufficiently high to decompose the carbonizable gas or liquid at those points and deposit on them a firm coating of carbon of good electric conducting power. As soon as this is done the filament ceases to glow. The current is now increased and the portions of next highest resistance have carbon deposited over them. In this manner, by gradually increasing the strength of the electric current, the entire filament glows uniformly under the influence of the current. The process generally requires but a few seconds. The flashing process not only renders the filament electrically uniform throughout its entire length, but also gives it a radiating surface better suited for use in a lamp.

The mounted and treated filaments are now to be sealed in an enclosing chamber of glass. For this purpose a glass globe G, of the form shown in Fig. 85, is employed, open at both ends at A, and B. The opening B, is sufficiently large to admit the mounted filament as shown in Fig. 83, as far as the shoulder S, when the two are then sealed together by the fusing of the glass as shown.

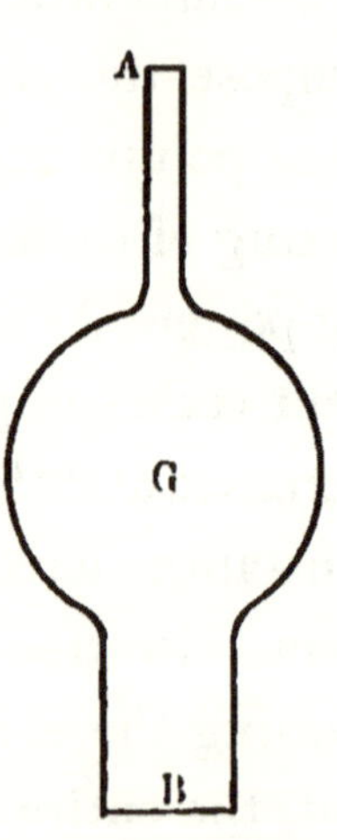

Fig. 85. Glass lamp-globe.

There now remains only the *exhaustion* of the lamp chamber, in order to complete the lamp. This is accomplished by means of air pumps. A number of lamps are connected to the same pump. In most cases the exhaustion is commenced by some form of good mechanical air pump, in which the valves are automatically opened and closed by the to-and-fro motions of the piston, the balance of the operation being completed by a mercury pump, which is capable of producing a much higher vacuum than the mechanical pump.

Mercurial pumps are of two types; viz., the *Geissler pump* and the *Sprengel pump*. In the Geissler pump advantage is taken of the Torricellian vacuum, which is found in the top of a dry glass tube over 33 inches in length, closed at one end, that has been filled with mercury and inverted, with its lower end dipping into the mercury. The atmospheric pressure at the level of the sea is only capable of sustaining a column of mercury of about 30 inches in vertical height, so that the mercury will run out of the tube until the mercury in the tube is about 30 inches above the level into which it is dipped. By means of suitable devices, the tube is successively filled and emptied of mercury and the lamps exhausted. A form of Geissler pump is shown in Fig. 86. The glass tube is provided at its upper extremity with a globe A, terminating with a 3-way tap C. In the position of the tap shown in the figure, A, is placed in connection with the tube *d*, and when turned through a quarter turn it is placed in connection with *e*. A reservoir B, filled with mercury, is connected with the lower end of the vertical tube by means of a stout rub-

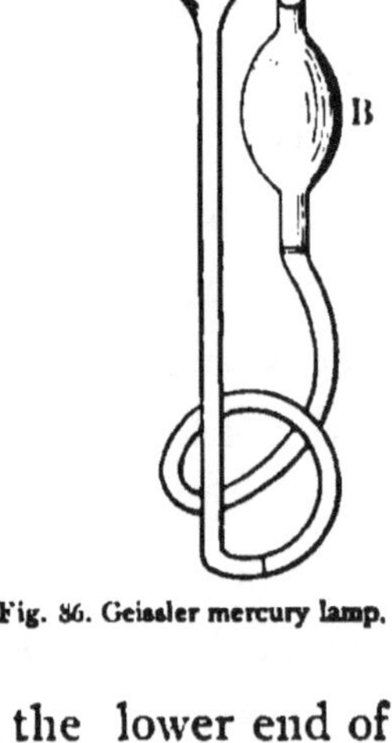

Fig. 86. Geissler mercury lamp.

ber tube. When B, is raised above A, the mercury rises and fills both tube and globe A. The tap is then closed and B, is lowered; the mercury now falls in A, until it stands at the barometric height, so that a Torricellian vacuum is left in A. C, is now turned so as to place A, in connection with d, to which a lamp or a row of lamps is connected, and the air in the lamp chamber expands and fills A. C, is now turned, B, raised, and the air in A, expelled through e, into the air and the operation repeated. In practice, devices are employed for mechanically filling and emptying the tube.

The Sprengel pump operates by means of a stream of mercury, which, in falling through a vertical tube, entangles bubbles of air which are thus removed from the vessel that is to be exhausted. A simple form of Sprengel pump is shown in Fig. 87. A vertical tube A, is provided with an enlargement at B, in which a lateral opening is provided at a, for the attachment of the chamber to be exhausted. In the upper part of B, is placed a tube *b*, shaped and connected as shown. A reservoir containing mercury is connected with *d*, by means of a stout rubber tube not shown in the figure. If this reservoir is raised above the level of *b*, the mercury rises in D, and passes through C, where it is freed from mechanical impurities. At this moment the tap is opened for a while and the air originally contained in D and C, expelled. *c*, is now closed and the mercury permitted to fall through B, into A. In so doing the air in the lamp chamber connected with *a*, is entangled with and gradually removed

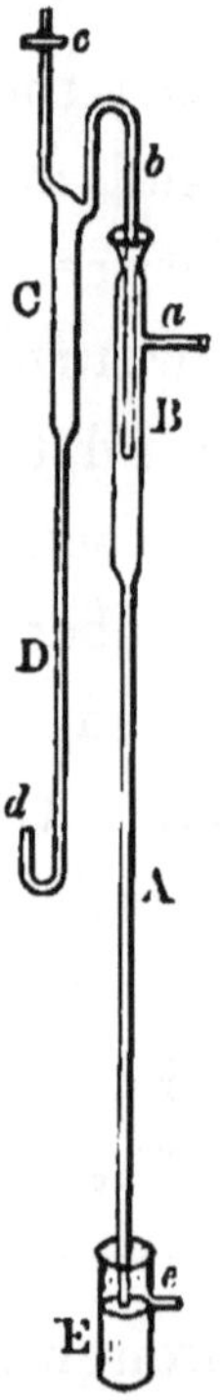

Fig. 87. Sprengel mercury lamp.

from it. In actual practice the movements are mechanically effected.

It is usual to obtain a very high vacuum in the chamber of an incandescent lamp. At one time it was usual to remove, say 999,999 parts of air in 1,000,000, or, to obtain a vacuum of the millionth of an atmosphere.

When the proper degree of exhaustion has been obtained, the next step is to *seal off* the lamp, which is effected by fusing the tube A, Fig. 85, at the top of the lamp chamber. In the early days of the manufacture of incandescent lamps this was done while the lamp was cold. The result was that the lamps failed after a comparatively short life. The reason is now easy to see. Carbon possesses in a remarkable degree the power of absorbing and occluding gases in its pores. It is this power that makes powdered carbon so effectual a disinfectant. Moreover, the walls of the lamp chamber have a film of condensed air coating them, and this remains even after the vacuum has been obtained. When, therefore, a lamp is turned on, the heat drives off the gas both from the carbon filament, and from the walls of the chamber, and thus the vacuum is ruined. If, however, before the lamp is sealed off, when a good vacuum has been obtained, a strong current is passed through the filament while the pumps are still operating, the heat drives off the occluded gas in the filament and the coating of condensed gas on the walls of the chamber, which is then carried off by the pumps. When, now, the lamp is finally sealed off, the vacuum it contains is not only high but will continue until it is spoiled.

CHAPTER VIII.

HOW THE ELECTRIC CURRENT SUPPLIED TO THE HOUSE IS MEASURED.

WE have seen how the energy of the burning coal in the central station is transformed into electric energy, and how, in its turn, this energy is transmitted through the street mains to the house, where it is transformed into the light and heat which appear in the lamp. This electricity is sold to the consumer by the Company that operates the central station. It, therefore, becomes necessary to adopt some means whereby the exact amount of electricity which passes into the house can be measured.

As is well known, the quantity of gas supplied to a house for lighting purposes is measured by the gas meter. The gas meter measures in cubic feet, the quantity of gas which enters the house. Without going into a detailed description of a gas meter, it will suffice to say that, as soon as a certain volume of gas has entered the meter, it is discharged into the house, and that each discharge produces a motion, which is recorded on a tell-tale dial.

Water meters have been constructed on similar principles, so that the actual quantity which passes into a house is recorded. In most cities, however, water meters are not used for measuring the consumption in private houses. It has been found easier to base the yearly charge at a certain rate per outlet or faucet, this rate being practically based on the mean quantity discharged per faucet. It will be understood that this mean quantity can be readily estimated by reference to the total quantity discharged into a reservoir in a given time, and the total number of faucets supplied from this reservoir.

In a similar manner the quantity of electricity supplied by the central station to each consumer is either estimated by a given charge of so much per lamp, or as actually measured by some form of recording meter. We will, therefore, briefly examine some of the different meters in actual use.

In the well known forms of gas or water meters, we record the number of cubic feet of gas or water that pass through them. In the case of the electric meter we record either the total quantity of electric energy, or the total quantity of electricity that has passed.

A form of electric meter, called the *chemical meter*, depends for its operation on the ability of an electric current to decompose certain chemical substances. Since such a decomposition is called *elec-*

trolysis, the chemical meter is sometimes called an *electrolytic meter*.

Fig. 88, shows a form of chemical meter in very common use, called the *Edison meter*. It consists of two zinc plates,

Fig. 88. Edison electric meter.

which have been covered with an amalgam of zinc. These plates are placed in bottles filled with a solution of zinc sulphate. On the passage of the current, metallic zinc is deposited on the plate connected with the negative terminal, while an equal quantity of zinc is dissolved or removed from the plate connected with the positive terminal. The amount of current passing is measured by the amount deposited on the negative plate, which is determined by the increase in weight. Two strips of

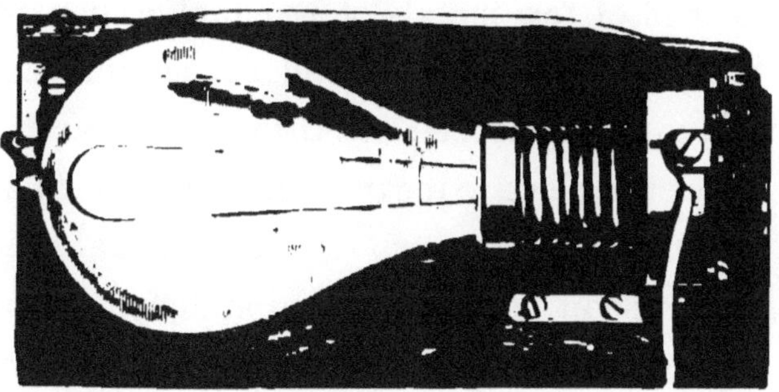

Fig. 89. Thermostat for Edison meter.

German silver R, R, provided with binding posts at P and N, for the attachment of the positive and negative supply wires, carry the greater part of the current, only a small but a definite fractional part passing through the meter bottles.

Where the meter is located in places where its liquid is liable to be frozen, it is provided with a *thermostat*, Fig. 89, which acts to keep it above a certain temperature. The circuit of the lamp is so arranged that as soon as the temperature falls below a certain predetermined point, the unequal expansion of a bar, formed of two metals of unequal expansibility, curves the bar and causes its free end to complete a circuit through the lamp. Since the electric incandescent lamp forms an excellent heater, the air inside the meter soon becomes heated sufficiently to straighten the bar, and thus automatically cut the current from the lamp.

A form of meter called the *Thomson recording meter* is shown in Fig. 90. In this meter the current passing, or, more correctly, the amount of energy which passes, is determined by recording the number of rotations made by a small electric motor placed in the circuit of the current to be measured. The circuit connections are such that the number of revolu-

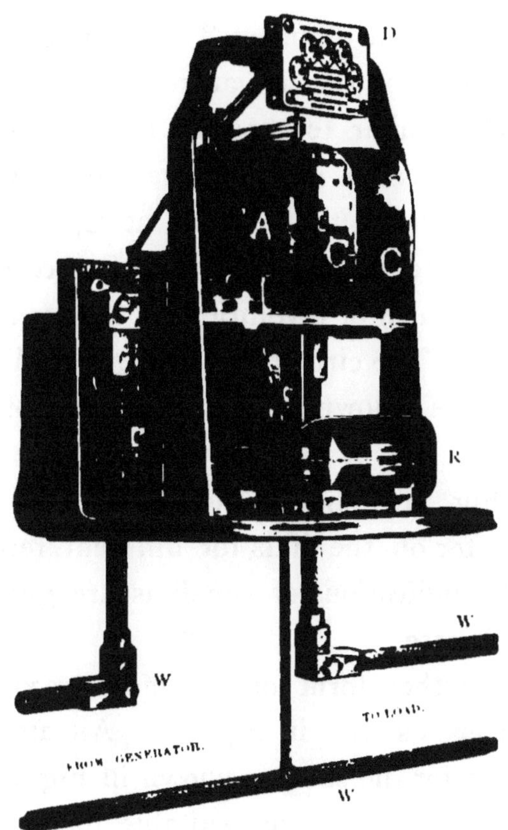

Fig. 90. Thomson two-wire meter.

tions of the armature depends on the amount of electric energy which passes through the circuit. The number of revolutions of the armature is recorded on dials similar to those on the gas meters. An inspection of the figure will show coils which carry the current at C, C. The armature is at A, and dials for recording the number of revolutions of the armature, at D. The circuit connections to W, W, W, are shown in the lower part of the figure.

Fig. 91, shows the position of the pointer on the dials for different records. The indications of the dials are given in each case.

Another form of Thomson recording meter is shown in Fig. 92. An air-tight cover for the same is shown in Fig. 93.

A form of meter suitable for use on arc-lamp circuits is shown in Fig. 94. This form of meter is similar to those shown in Figs. 90 and 92.

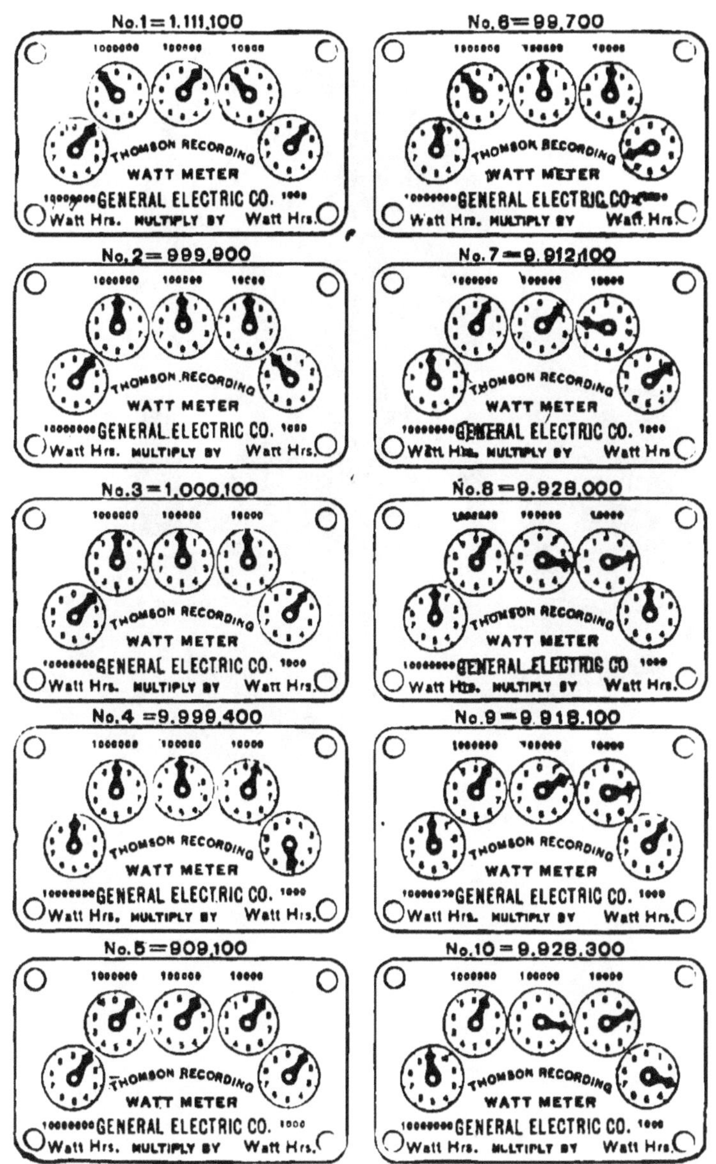

Fig. 91. Records made on Thomson recording watt-meter.

Fig. 92. Thomson recording watt-meter.

Fig. 91. Thomson recording meter with cover.

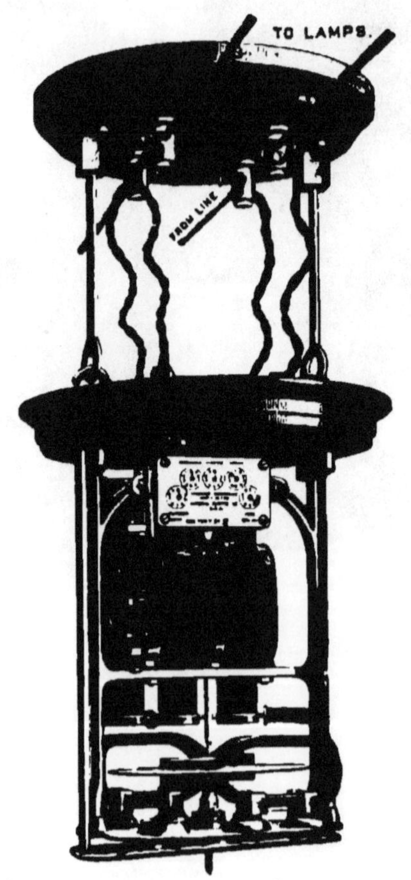

Fig. 94. General Electric Company's arc-circuit meter

CHAPTER IX.

HOW THE ARC LAMP OPERATES.

As we pass through the streets, tracing in the imagination the course of the underground conductors which connect the central station with the wires in the houses, we will probably notice, hanging over the streets from supports, or mounted on poles, large electric lights resembling that shown in Fig. 95. These are called electric *arc lamps*. If we come near enough to one of these lights, we, probably, see something like what is shown in Fig. 96. Arc

Fig. 95. Electric arc lamp.

lamps give a much greater amount of light than do incandescent lamps, the ordinary arc lamps employed for lighting streets, or for open areas generally, are commonly rated as furnishing an intensity of light equal to that of 2,000 ordinary candles.

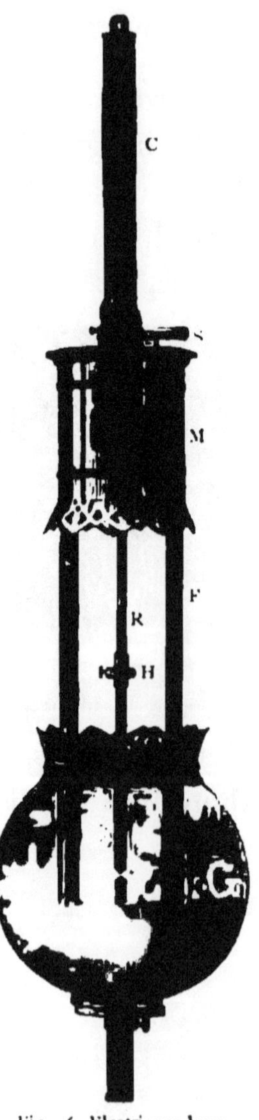

Fig. 96. Electric arc lamp.

When two carbon rods or pencils are placed in contact in the circuit of a sufficiently powerful electric source; such, for example, as a dynamo, and are separated from each other, a brilliant arc or bow-shaped mass of light called the *voltaic arc*, or the *electric arc*, is formed between them. The ends of the carbons become intensely heated, and throw off or emit the well known light of the electric arc-lamp. It is between two such carbon rods P and N, that the light is produced in the arc lamps referred to in Figs. 95 and 96.

The voltaic arc forms the most intense source of artificial heat known. All the metals, even platinum, melt readily when brought into it. Carbon, one of the most refractory substances known, is volatilized in its intense heat. It is because the carbon electrodes or rods between which the electric arc is formed, are raised to so high a temperature that they emit so bright a light.

A voltaic arc is too bright to examine with the unaided eye, but if we place a lens before a carbon voltaic arc, so as to form an image of the arc and the electrodes, on a distant screen, we will see something like what is shown in Fig. 97.

Examining this attentively, we will see that a tiny *crater* has been formed on the end, P, of one of the carbons and that a tiny *nipple* or *projection* has been formed on the opposing end, N, of the other carbon.

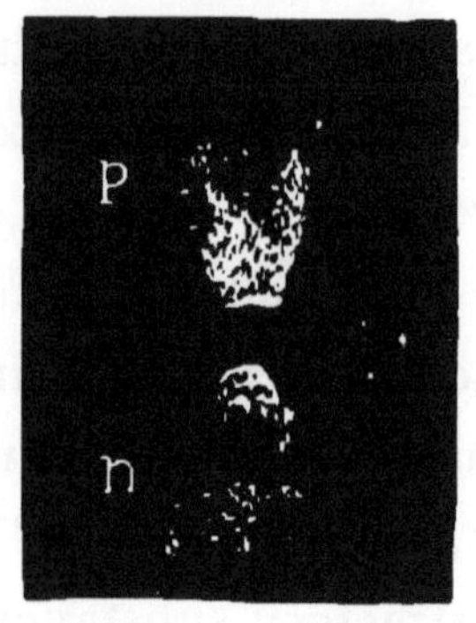

It is not difficult to understand how these changes have been produced in the ends of the carbons during the establishment of the voltaic arc between them. Under the intense heat of the arc, the carbon connected with the positive pole of the source, that is, the *positive carbon*, has been volatilized; i. e., reduced to a state of a vapor, so that its end has been hol-

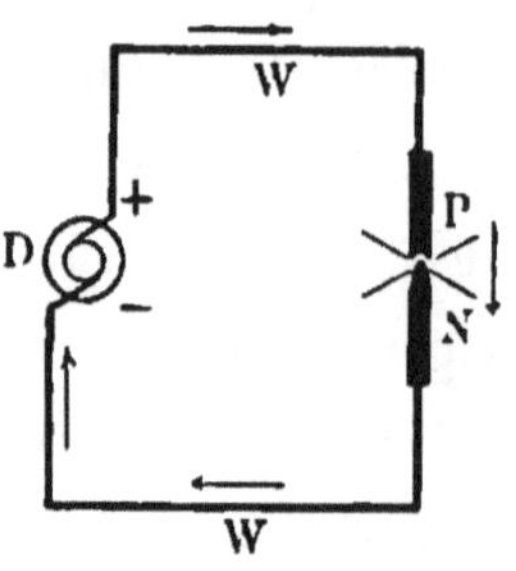

Fig. 97. Voltaic arc.

lowed out. This carbon vapor has then been deposited in a solid form on the opposing end of the carbon connected with the negative pole of the source, or the *negative carbon.* It is evident then, since carbon is fused or volatilized, that the temperature of the arc is extremely high. Moreover, since the carbon is deposited on the end of the negative rod, this rod must be much cooler than the positive rod. The positive carbon is therefore the principal source of light of the arc lamp. For this reason, where, as is generally the case, it is desired to throw the light downwards, the positive carbon is always made the upper carbon in the arc lamp.

Under the action of the intense heat, a curious change is produced in the carbon electrodes near their ends; viz., they are changed into a variety of carbon known as *graphite* or *black lead*, a substance well known for its use in lead pencils. There will generally be found in the neighborhood of the arc lamps in streets, small pieces of the electrodes which have been thrown away by the lamp trimmer, when re-carboning the lamps. If some of these are collected it will not be difficult to determine which were the positive carbons and which the negative. Moreover, it will be found that, whether positive or negative, the burned ends can be used for lead pencils, the graphite serving excellently for this purpose, although the rest of the carbon is too hard to leave any mark on paper.

The voltaic arc is also formed between metallic substances. In this case the metal is volatilized. *Metallic arcs* are colored by the volatilized metals. Copper forms a green arc. A copper arc is frequently seen on trolley lines, when the trolley wheel is momentarily jarred away from the trolley wire. It appears as a brilliant flash of green light.

If, during the burning of a carbon voltaic arc, the electrodes are examined through a piece of smoked or colored glass, it will be observed that the position of the crater in the positive carbon does not remain fixed, but shifts or moves from place to place. This is due to the fact that during the maintenance of the arc, the carbons are consumed or waste away, and the arc tends to establish itself between those points on the electrodes which are the nearest together. It is also due to the fact that it is difficult to obtain carbons that are of the same composition and density throughout, so that the arc tends to form at places where the carbons are most readily volatilized. This shifting of the position of the crater produces the well known flickering or unsteadiness of the arc light.

Since the carbons are consumed during the establishment of the arc, it is necessary to employ some means whereby they can be kept at a constant distance apart. This is done by means of various forms of *arc-light mechanisms.*

Examining the arc lamp shown in Fig. 96, it will be seen that the two carbon rods P and N, are placed vertically one above the other. The upper carbon is attached to a lamp rod R, by means of a carbon holder H. The lamp chimney C, is provided to receive the lamp rod. M, contains the mechanism provided for keeping the carbons at a constant distance apart, during the operation of the lamp. Arc-lamp mechanisms are of various forms. They consist generally, however, of a *gripping device* for taking hold of the lamp rod, raising it, and establishing the arc between them, and for releasing the lamp rod, and thus permitting the upper carbon to fall towards the lower carbon, when, by combustion, the carbons are too far apart.

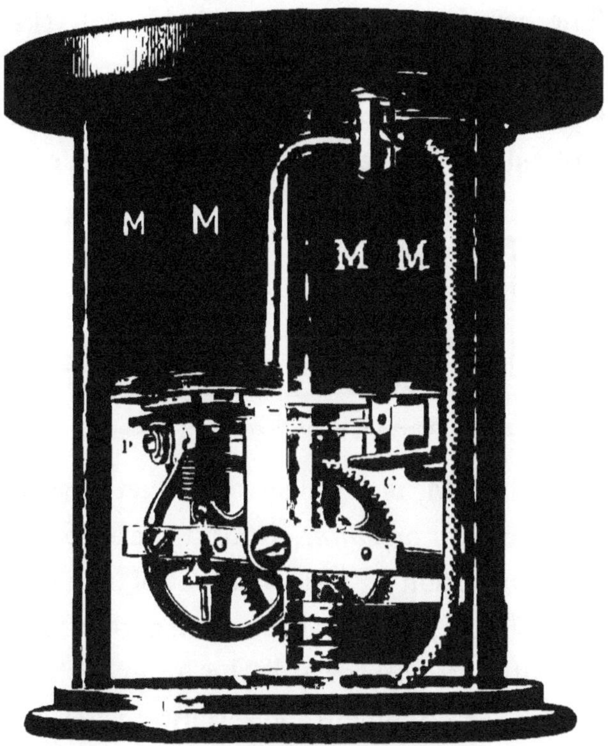

Fig. 98. Mechanism of arc lamp.

A form of lamp mechanism for feeding the carbons is shown in Fig. 98. M M, is an *electro-magnet* ; i. e., a device in which the magnetism can be turned on or off by turning the current on or off the coils. The lighting current passes through the coil M M, and the attraction by these coils of the iron core C, lifts the carbon rod and establishes the arc between the car-

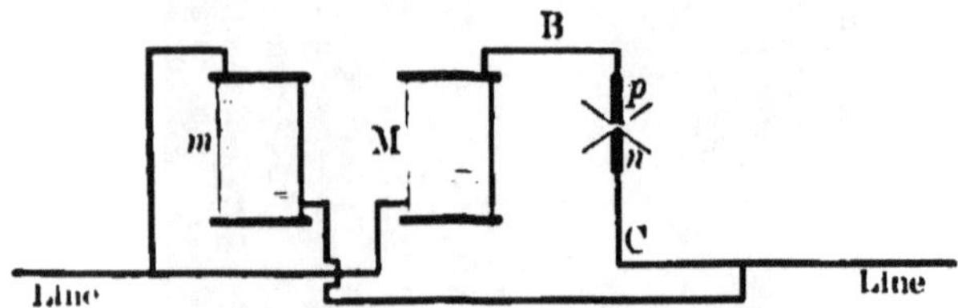

Fig. 99. Diagram of magnet in arc-lamp mechanism.

bons. M M, is an electro-magnet placed in a by-path or, *shunt*, around the carbons. The attraction by this magnet of the soft iron plate P, releases the upper carbon and permits it to fall towards the lower carbon.

The arrangement of circuits whereby the carbons are automatically kept at a constant distance apart is shown in Fig. 99. The magnet M, whose coils are formed of short thick wires, and whose electric resistance is, therefore small, is placed in the same circuit with the arc as shown. The magnet m, whose coils are formed of long, thin wire, and whose electric resistance is, therefore, large, is placed in a by-path around the magnet M, and the arc. Consequently, nearly all the current passes through the magnet M, and the arc, very little passing through the magnet m. But, as, by consumption, the carbons get a greater distance apart, the resistance of the arc circuit MBC, increases, so that a stronger current passes through m, until finally the attraction of its armature becomes sufficiently great to permit the upper carbon to feed or drop downwards to the lower carbon. This alters the resistance of the branch MBC, and permits the magnet M, to again attract its armature and again separate the carbons.

As we have already seen, arc lamps are connected to the circuit *in series;* i. e., the current passes successively through all the lamps, as shown in Fig. 30. Consequently, if any lamp failed to operate, the circuit would be broken and all the

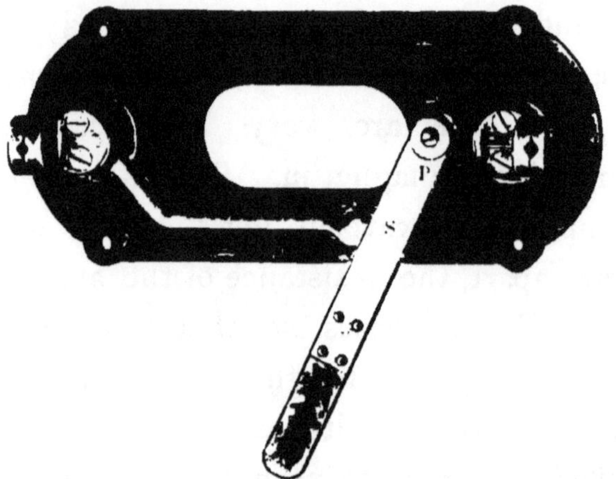

Fig. 100. Arc-lamp switch on stationary lamp hanger.

lamps would be extinguished. This is prevented by means of a device which automatically establishes a by-path of small resistance past any faulty lamp, thus preventing the circuit from being accidentally broken or opened.

Each lamp is provided with a *short-circuiting switch*, whereby the lamp may be cut out from the circuit without affecting

Fig. 101. Arc-lamp switch.

the other lamps. Such a switch is shown in Fig. 96, at S. Its general construction is shown in Fig. 100. The line circuit L L, can be connected by means of a metallic lever S, pivoted at p, by bringing the switch into the position shown in the figure. Under these circumstances, so much of the current passes through the short circuit L S L, that the lamp is practically cut out from the circuit. Another form of switch suitable for arc lamps is shown in Fig. 101.

The proper operation of the arc lamp requires the carbons to be placed one over the other in exactly the same vertical line, since if they are much out of line the proper feeding of the upper carbon would be prevented, the carbon in falling making a side contact with the lower. Therefore, the lower carbon holders are arranged so as to permit a slight lateral displacement of the lower carbon, and thus to bring it directly under the upper. A few forms of carbon holders are shown in Fig. 102. Those at A and B, are intended for the

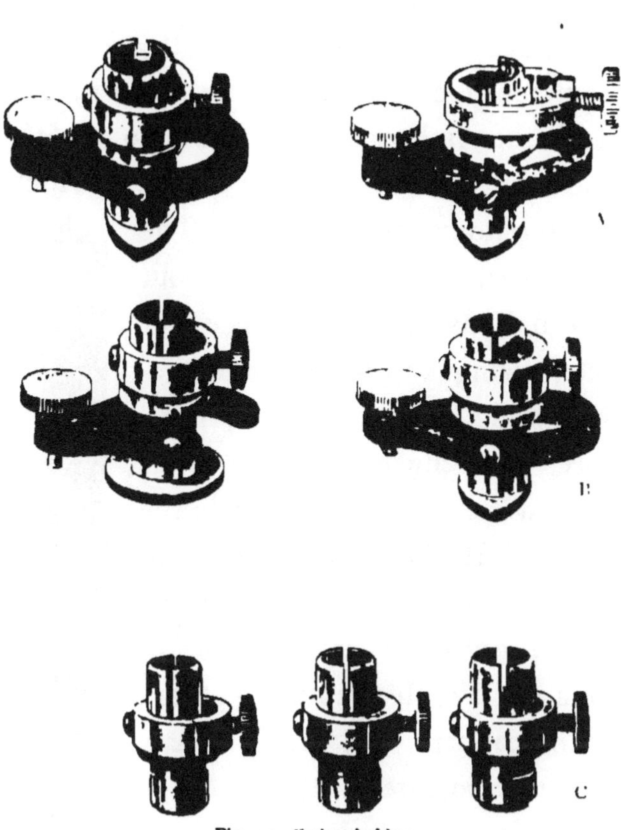

Fig. 102. Carbon holders

lower carbons, and those at C, for the upper carbons.

The arc-lamp carbons that are most commonly employed in this country are 12 inches in length, and from ½ to 7–16th inch in diameter. Their life will be from 7 to 9 hours, respectively. The lower carbon only consumes about one half as rapidly as the upper one, owing to its lower temperature and deposit of carbon; therefore, the length of the lower carbon is only half as great as the upper. It has been found, in practice, that the length and life of the ordinary carbons is increased by electroplating them with copper. An arc-light carbon, electroplated with copper at *a*, *a*, is shown in Fig. 103.

Fig. 103. Solid carbon.

In order to decrease the unsteadiness of arc lights due to the travelling of the arc, what are called *cored carbons* are employed. These consist as shown in Fig. 104, of carbons provided with a cylindrical core of softer material at c, c. The effect of the central core is to maintain the position of the arc at the centre of the positive carbon.

c

c

Fig. 104. Cored carbons.

Where arc lamps are required to burn for longer periods than 7 or 9 hours, it is necessary to *re-carbon* them during their operation. For this purpose the lamp switch is closed, and new carbons are put in place. In order to avoid the expense of re-carboning at night, as well as the danger of working on a live circuit of high voltage, a device called a *double-carbon* or *all-night arc lamp* is generally employed. This consists as shown in Fig. 105, of a lamp provided with two pairs of carbons A A, B B, so arranged that when one set is consumed the current is automatically transferred to the other set. Separate lamp chimneys at C, C, are provided for the lamp rods. By means of this device the lamp can readily be made to burn for from 14 to 18 hours, according to the size of the carbons and the current passing through them.

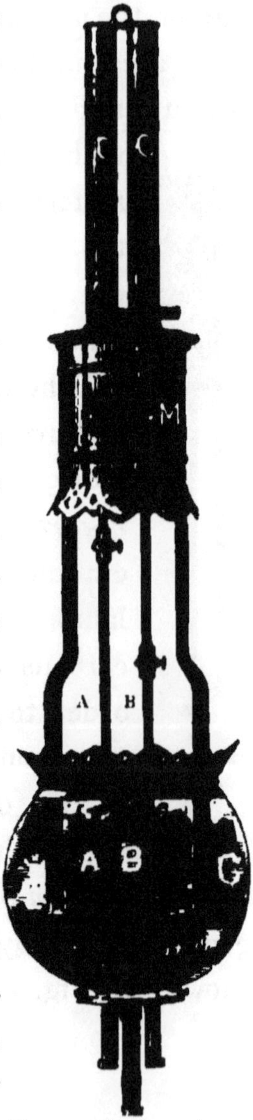

Fig. 105. All-night arc lamp.

In the arc-lamp mechanisms so far shown, only the upper carbon is fed, the lower carbon remaining fixed. Consequently, the position of the arc is gradually lowered as the lamp burns. This is a matter of no consequence where the lamp is employed for general lighting purposes. But where the light is to be placed at the focus of a mirror, as in the *theatre arc lamp* shown in Fig. 106, or in the *locomotive head light* shown in Fig. 107, such a lamp would be impracticable, since as the carbons consumed, the arc would be removed out of the focus of the mirror. In such cases it is necessary for the lamp mechanism to feed both carbons, the lower

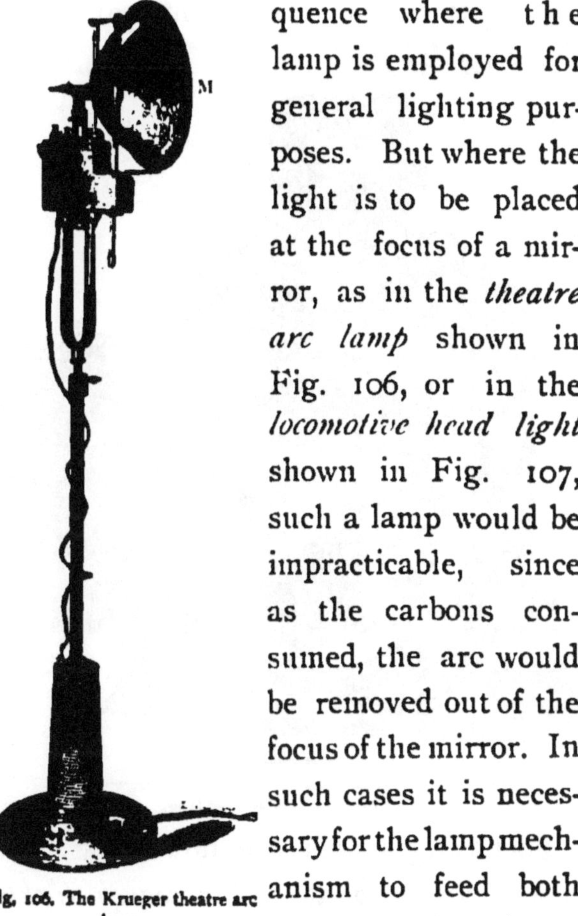

Fig. 106. The Krueger theatre arc lamp.

carbon being fed at about half the rate of the upper carbon. Such lamps are called *focussing lamps.*

Fig. 107. Locomotive head-lamp.

Focussing lamps are now very commonly employed at sea for *search lights.* A search light, or *naval projector*, is shown in Fig. 108. The positive carbon p, is shown about twice the length of the negative carbon. The carbons are placed so as to bring the arc at the focus of the reflector R. By this means a powerful nearly parallel beam of light is produced which may be thrown a considerable distance without suffering much decrease in intensity. A smaller search light is shown in Fig. 109.

In most central stations for lighting purposes, arc lights are distributed as well as incandescent lamps. Arc-light dynamos do not differ essentially from incandescent dynamos. Since, in most cases, a large number of arc lamps are connected in series, the total pressure or E. M. F., which it is necessary to produce, is very high. Each ordinary carbon arc requires a pressure of about 50 volts to be maintained at its terminals. Arc-light dynamos are now constructed which operate as

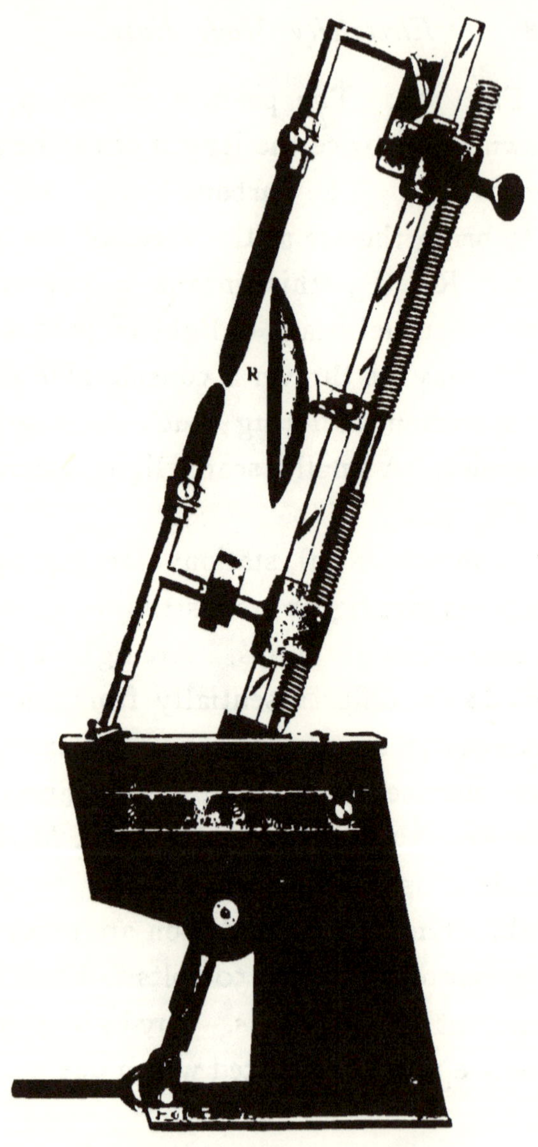

Fig. 108. Focussing arc lamp for naval projector.

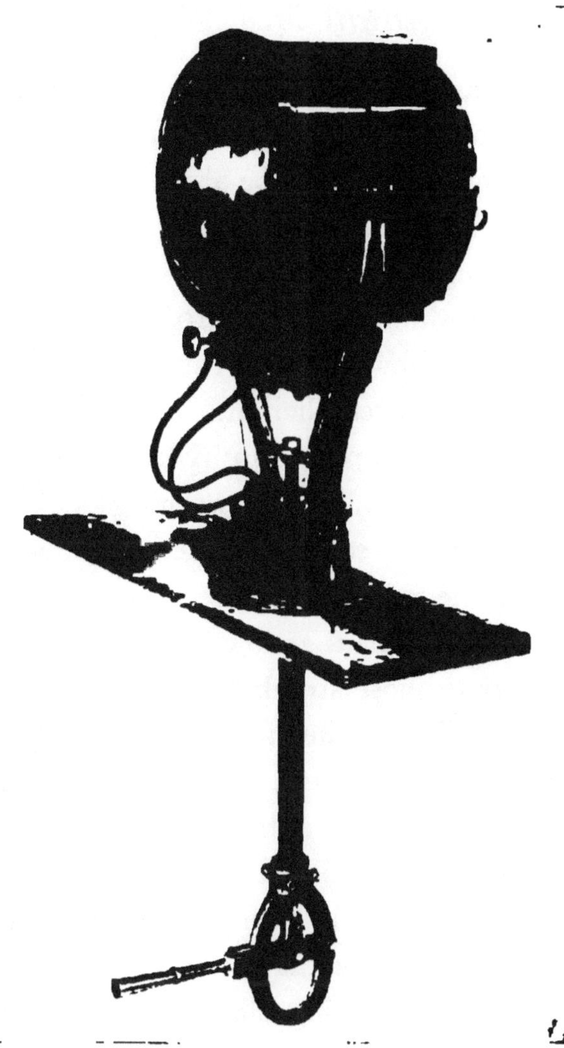

Fig. [illegible]. Small headlight.

many as 120 lamps in a single circuit; the total pressure required in such a circuit would, therefore, be about 6,000 volts. This high pressure would correspond to a water motive force or pressure such as shown in Fig. 110, capable of throwing a stream of water to a great height. Such streams are employed in hydraulic or placer mining, for washing out beds of earth in order to obtain the gold. They are sometimes so powerful as to cause accidents to those struck by them. In a similar manner high electric pressures, as for example those used in some arc-lighting circuits are very dangerous. It is necessary, therefore, to be careful not to handle incautiously the conductors that carry arc light currents. In lightning flashes the electric pressure or E. M. F. must be exceedingly

Fig. 110. A high water motor force.

high, as may be inferred from the air space through which the discharge often passes. Such an electric pressure as will produce a lightning flash, or electric discharge, a mile in length is, perhaps, many millions of volts.

An arc-light dynamo capable of feeding 120, 2,000-candle power arc lamps, directly coupled to a driving engine, is shown in Fig. 112. In order to permit any of the lamps in the series-connected circuit to be cut out at will, the dynamo is necessarily provided with a device called an *automatic regulator*, by means of which the electric pressure of the dynamo, is varied, either by shifting the position of the collecting brushes on the commutator, or, by some other means.

Fig. 112. 120 2000-candle power Brush arc dynamo, coupled directly to a vertical compound engine.

CHAPTER X.

HOW THE LIGHT OF ELECTRIC LAMPS IS BEST DISTRIBUTED.

HAVING now briefly traced the manner in which the electric current is distributed, how it is produced, and how it is caused to produce light, in both incandescent and arc lamps, we will very briefly consider how the light can be best distributed for the various purposes for which it is employed. But before we do this, it will be necessary to give brief definitions of a few words that will be frequently used.

We should carefully distinguish between the *light* which a luminous source, such as a lamp produces, and the *illumination* it can effect. The word light refers to what is emitted by the source, and the word illumination to the effect produced by the light in falling upon a surface, that is to the quantity of light received per square inch or per square foot, either directly from the luminous source, or indirectly by reflection from neighboring bodies. Lamps are rated in candle-power. This standard is the *British standard candle*, a candle of known dimensions and composition, that burns 2 grains per minute.

The illumination of any surface will depend, both on the candle-power of the luminous source, from which it receives light, and also on its distance from this source. The greater the candle-power of the source, and the nearer the surface to be illuminated or lighted is to the source, the greater will be the illumination of that surface. In open areas, where there are no reflecting walls, the light that can be used for purposes of illumination is limited to that received directly from the source, but in a room, the surface to be lighted receives its light not only from the source, but also from all reflecting surfaces in the room. Therefore, the amount of light received, say on the surface of a desk or table, will not only depend on the number of lamps in the room, and on their situation, but also on the character of the walls and ceiling of the room, and on their ability to throw off light. As a rule, dark, dull, rough surfaces absorb so much of the light, that in some rooms but little light reaches the surface to be illumined, such, for example as a desk or table, except that which comes directly from the luminous source. This must be carefully borne in mind in determining the number of lamps that are to be placed in a room in order to obtain the desired illumination.

In the distribution of light for the purposes of illumination, shadows will be best avoided by dividing the light into as many small sources as possible. For this reason the arc lamp is not well suited for interior lighting, since its candle-power is so great that a single lamp gives off more than sufficient light for any but a large room. Incandescent lamps are, therefore, to be preferred for agreeable lighting since they can be distributed throughout the room.

Where a room is to be lighted by incandescent lamps, having determined the number of lamps required, it remains to consider how the lamps can best be placed for tasteful and effective illumination. They should not be massed too closely together, thus leaving portions of the room in comparative darkness. Clusters of lamps in electroliers may be used; or, the lamps may be separately placed, at more or less equal distances apart, near the ceiling, or in cavities in the ceilings or walls provided for the purpose. In any assembly room, where the audience always faces the same direction, the lamps will preferably be placed so as to throw the light directly on the stage or platform, and where they are shielded from the eyes of the audience. Where electroliers are used, reflectors may be employed; or, what is perhaps better, except in very large rooms, the lamps may be covered by ground glass shades or ground glass globes; for, although such globes or shades, cut off considerable light, yet, since the entire surface of the globe or shade becomes illumined the light is thus better scattered or diffused, and the general illumination greatly improved. Moreover, such globes may be made quite ornamental, as is shown in Fig. 113.

The arc lamp is best suited for out-of-door lighting. For electric light-house

Fig. 113.—Ornamental lamp globe.

purposes arc lamps only are used. In this case, since it is desired to throw a beam of light to a great distance, the lamp is provided with a cylindrical lens, in the shape of a lantern surrounding the lamp, so that slightly diverging rays are thrown to a great distance, as shown in Fig. 114.

Fig. 114. Electric lighthouse.

For the illumination of large open air areas, such as squares or public gardens or parks, the system called *tower-lighting* is frequently employed. In such cases a number of arc lamps are grouped together, and placed at the top of a tower, as shown in Fig. 115. Where a number of such electric light towers are employed, and the space to be lighted is sufficiently open, tower lighting produces an effect that closely resembles moonlight.

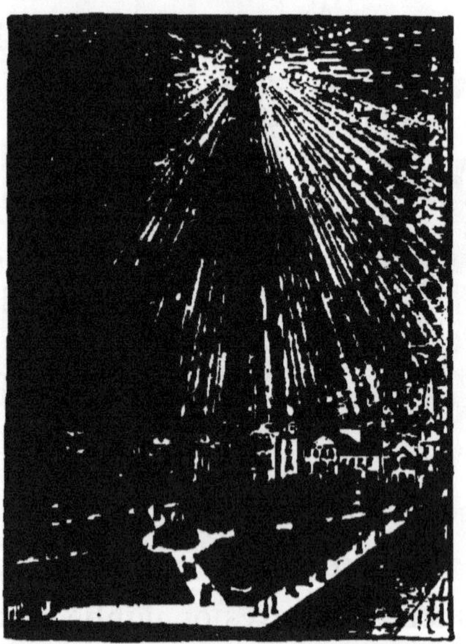

Fig. 115. System of electric town lighting.

For street lighting, single arc lamps are sometimes placed on the top of poles, such as shown in Fig. 116. Here a pole is provided with a number of *pole steps* s, s, s, etc., so as to permit the trimmer to readily get at the lamp to re-carbon it. Where it is necessary to place the lamp over the street, in order the better to illumine the road bed, the lamp is supended from the end of a *mast-arm*, A, supported on, the pole P P, as shown in Fig. 117.

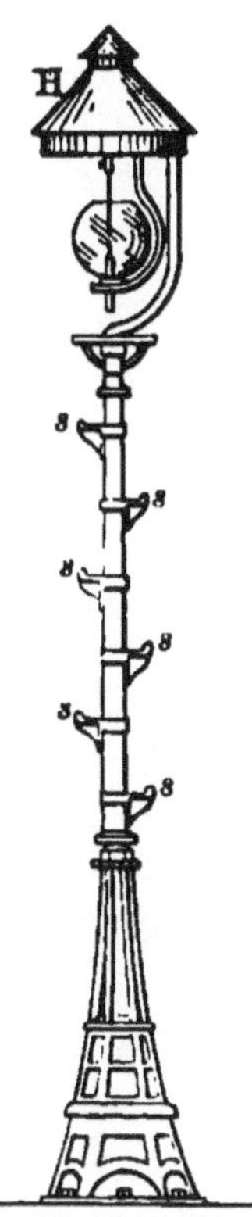

Fig. 116. Street lamp post.

In all cases where arc lamps are exposed to the weather, it is necessary, or at least advisable, to protect the lamp mechanism by a suitably shaped *hood*, such, for example, as is shown in Fig. 118. The inside of the hood is

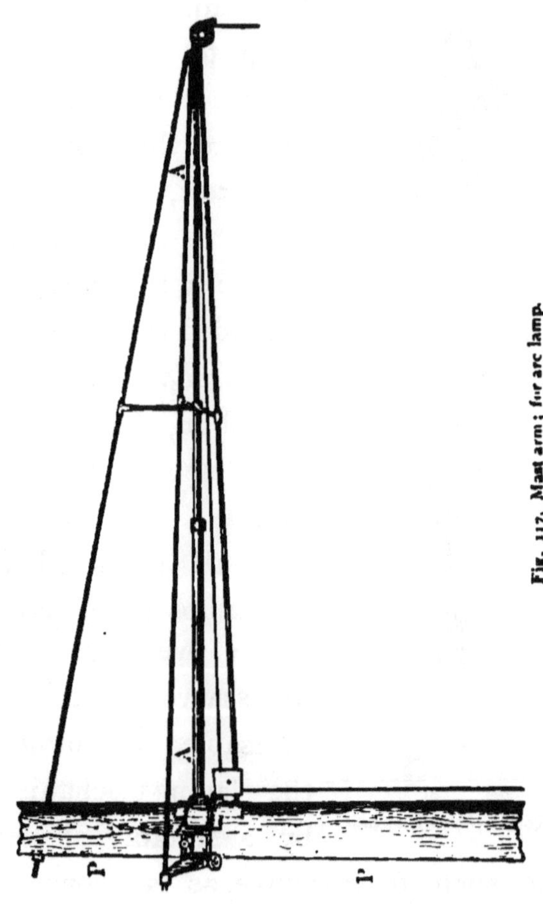

Fig. 117. Mast arm; for arc lamp.

smoothed and whitened, so that it may serve to throw the light downwards. Such hoods are usually placed on the pole in the manner shown in Fig. 119, which represents an *all-night* or *double-carbon*

Fig. 118. Lamp hood for arc lamp

lamp, inserted in the circuit between the wires W and W, and mounted on a pole P; or, as in Fig. 120, which represents what is called an *outrigger*, for suspending an arc lamp.

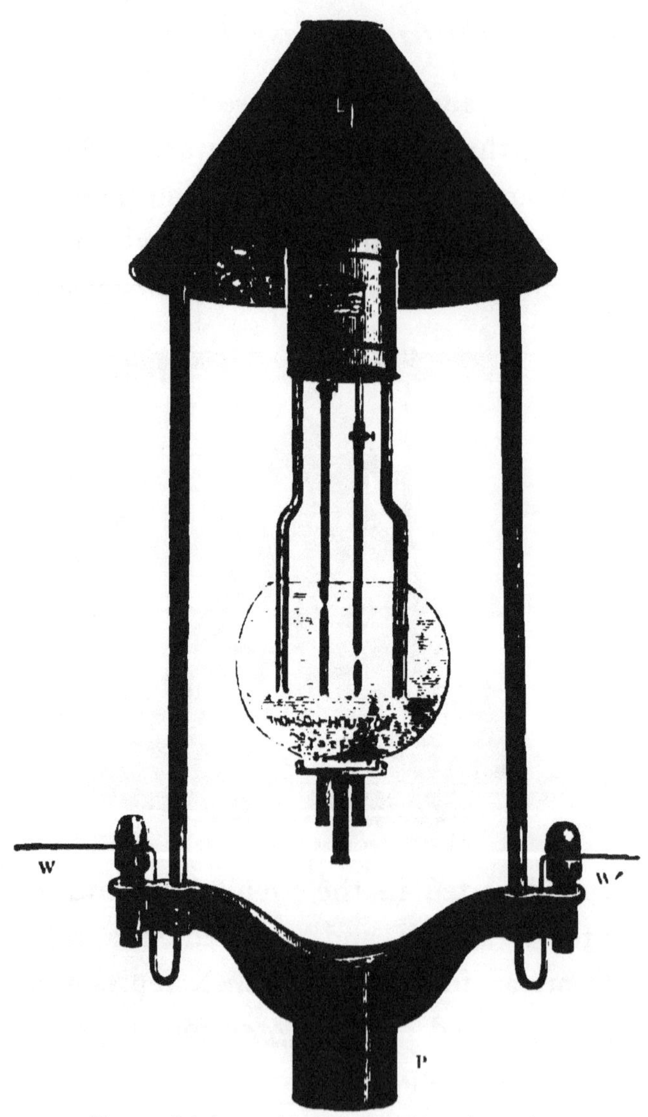

Fig. 119. Pole irons and hood, with double-carbon arc lamp.

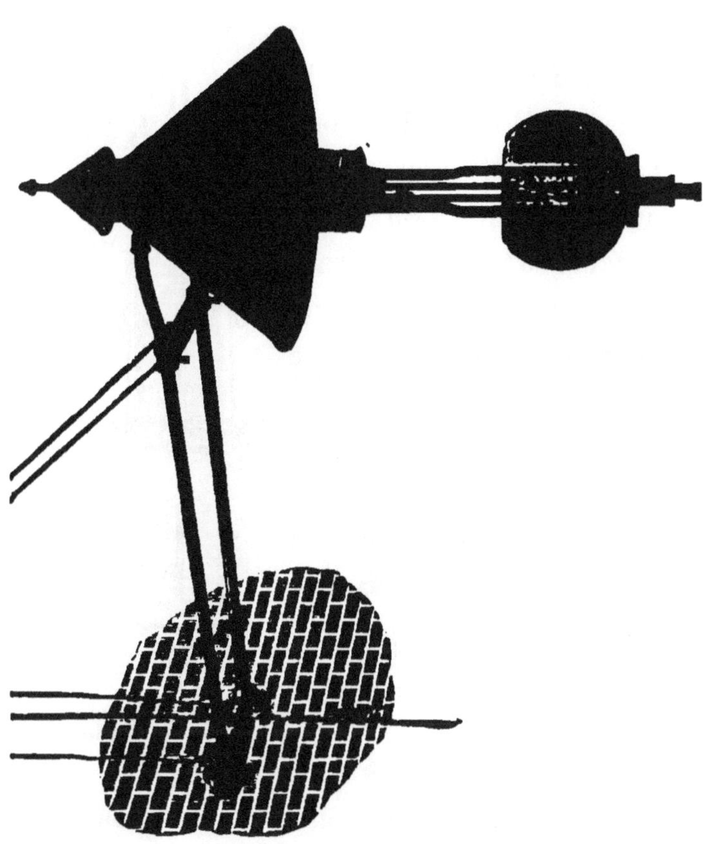

Fig. 120. Outrigger for suspending an arc lamp.

We have seen that incandescent lamps are almost always supplied with electricity in parallel, while arc lamps are usually supplied with electricity in series. When, therefore, it is desired to operate arc lamps in connection with a system of incandescent lamps, means must be provided for supplying the arc lamp in parallel with the incandescent lamps. This is done by connecting a suitable resistance of wire with the arc lamps, which are sometimes connected two in series across the mains.

Fig. 121. Arc lamp suitable for use in Incandescent lamp.

A lamp suitable for such use on incandescent circuits is shown in Fig. 121. Where arc lamps are to be used in rooms whose ceilings are low, it is necessary to adopt some means whereby the height of the lamp can be decreased. Such *short arc lamps* are

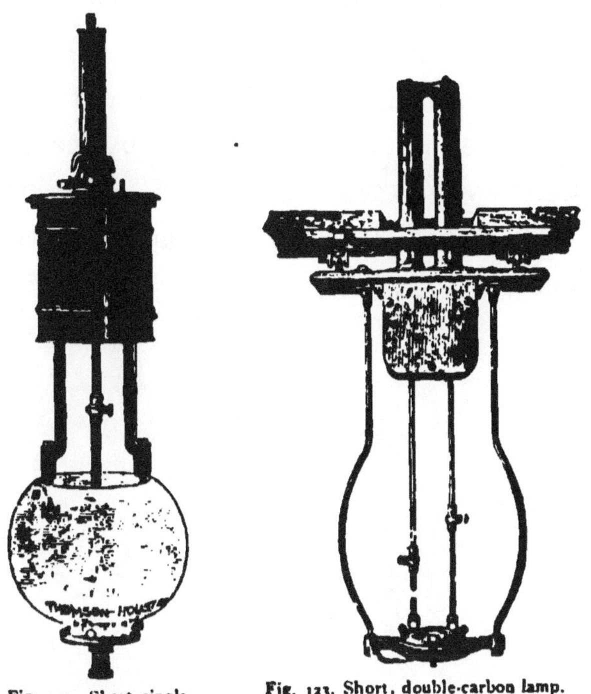

Fig. 122. Short single-carbon lamp.

Fig. 123. Short, double-carbon lamp.

shown in Figs. 122 and 123, of two different types, the former being a single-carbon lamp, and the latter a double-carbon lamp.

We have described incandescent and arc lamps as using *direct currents* only, that is currents that flow in one and the same direction. Both incandescent and arc lamps may also be operated by means of *alternating currents*; i. e., currents which flow alternately in opposite directions.

CHAPER XI.

THE VOLTAIC CELL AND HOW IT OPERATES.

IN 1786, the world was greatly astonished and excited by the announcement made by one Luigi Galvani, an Italian physicist, that he had discovered the cause of vitality or life. Happening to hang the legs of recently killed frogs, which he had prepared for certain electrical experiments, against the iron railing of a balcony, he was surprised to see them go through convulsive movements as in life. Galvani believed that he had discovered the cause of life, and quite naturally this announcement caused his experiments to be repeated all over the world. It was not long, however, before Alexander

Volta proved that what Galvani had discovered was not the cause of vitality, but a new method for producing electricity; i. e., a new electric source. He showed that the movements of the frog's legs took place more readily when two dissimilar metals, such, for example, as copper and zinc, were placed on a nerve and muscle respectively, and their free ends brought into contact. This experiment can be easily repeated by taking the hind legs of a recently killed frog, exposing the nerves on each side of the vertebral column, which appear as white threads, placing a strip of zinc Z, in contact with the nerve, and a strip of copper C, in contact with the muscle, as shown in Fig. 124. As soon as contact is made between the two pieces of metal, the legs will be convulsed as in life.

Volta's investigations of Galvani's experiments soon led to his invention of the *voltaic pile.* This invention may justly be regarded as the most important discovery made in electrical science up to that time, since it placed in the hands of scientific men means for readily produc-

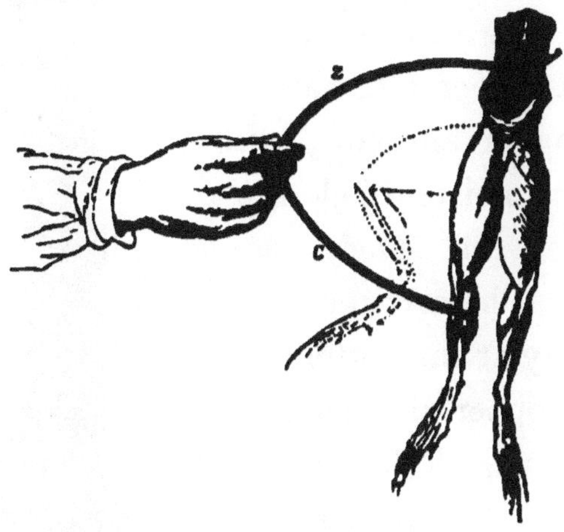

Fig. 124. Galvani's experiments.

ing electric discharges or currents. Volta believed that the electricity in his pile was due to the mere contact of dissimilar metals, but it was soon discovered that no continuous current was produced unless a chemical action took place. The voltaic pile is sometimes called the *Galvanic pile*. This, however, is erroneous, since Volta and not Galvani, was its inventor.

One of the earliest forms of Volta's piles or batteries is shown in Fig. 125. It consisted of alternate discs of copper and zinc, with a disc of moistened cloth placed between every second pair. The arrangement being say at the bottom of the pile, copper, zinc, cloth; copper, zinc, cloth, etc., until the pile was completed. The positive and negative terminals of the pile were situated at C and Z, at the bottom and top respectively.

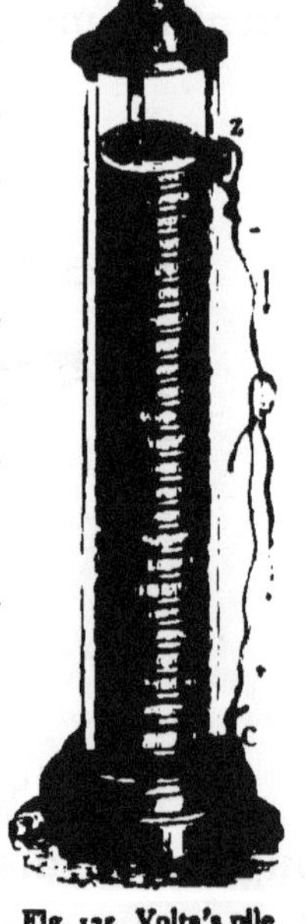

Fig. 125. Volta's pile.

Volta's original pile or battery has been greatly improved since its first invention. It is now generally made in what is called a single *voltaic cell*, a number of these cells being connected together to form a *voltaic battery*. Very many different forms of voltaic cells have been made, but all consist of what is called a *voltaic pair* or *couple*, consisting of two dissimilar substances, generally metals, immersed in a liquid substance called an *electrolyte*, capable of chemical action on one of the substances in the pair. The substances most frequently employed for a voltaic couple are zinc and copper, or zinc and carbon.

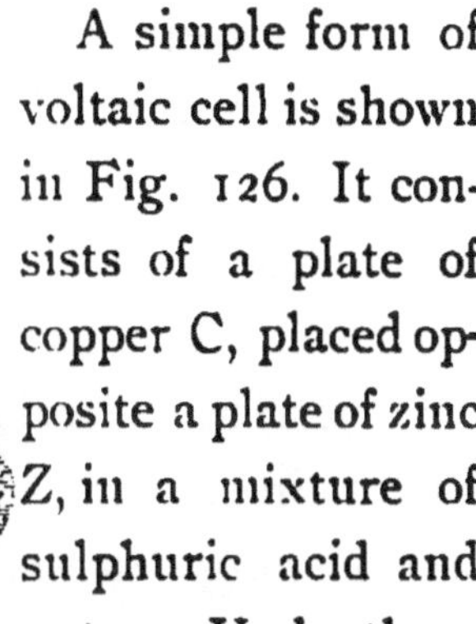

Fig. 126. Simple voltaic cell.

A simple form of voltaic cell is shown in Fig. 126. It consists of a plate of copper C, placed opposite a plate of zinc Z, in a mixture of sulphuric acid and water. Under these circumstances a chemical action takes place on the zinc, and an electric current flows through the circuit in the direction indicated by the arrows.

When a plate of zinc and a plate of copper are placed in a weak solution of sulphuric acid and water, without touching one another, the zinc, if pure, will not be attacked, and no chemical action will take place. If, however, the two plates are connected by metallic wires, as shown in Fig. 126, the zinc commences to be dissolved, and gas is now given off from the surface of the copper plate. The liquid is also heated, and an electric current is produced which flows through the circuit.

We saw, in the case of the central station, that the energy necessary to produce the electricity that flowed through the conductors in the streets, to operate the incandescent lamps in the house, came from the coal which was burned under the boilers. The question naturally arises, what is the source of the energy which produces the electricity in the voltaic cell? Remembering the chemical action, or the burning or dissolving of the zinc in the acid, when the electricity was produced, it would seem that the energy which produces the electricity in the voltaic cell is the energy produced by the burning of the zinc in the acid liquid or electrolyte; and this assumption is correct.

We have spoken of the voltaic cell as producing electricity. In reality, like all other electric sources, it does not produce electricity, but an E. M. F., or electric pressure which, when permitted to act on a circuit, produces an electric current in that circuit. In the case of any voltaic cell the length of time during which its current can be maintained is dependent on the quantity of the zinc and of the exciting liquid or electrolyte present.

When the circuit of such a voltaic cell is first closed the electric current it gives is the greatest. It is not long, however, before the current grows considerably weaker. This is mainly due to what is called the *polarization* of the cell. During action, one of the plates becomes covered with gas which tends to decrease the electric pressure produced by the cell, and hence the strength of the current it supplies.

Various methods are adopted for preventing the polarization of voltaic cells. Probably, that in most common use is to surround the plate, on which gas collects, by a liquid capable of either dissolving the gas or of preventing it from forming. In such cases there will be two liquid substances or electrolytes, and the cell becomes what is called a *double-fluid cell.*

Ordinary zinc will readily dissolve when placed in dilute sulphuric acid, so that the cell would waste away on open circuit. In order to prevent this the zinc is *amalgamated*, or coated with an amalgam of zinc and mercury. This is readily accomplished by dipping the zinc in dilute sulphuric acid, and then rubbing a small quantity of mercury over it. An amalgamated zinc has a bright, shiny surface, and feels greasy to the touch.

The pole or terminal of the voltaic cell from which the electric current is assumed to pass out, is, as in the case of any electric source, called the *positive terminal* or *pole*, while that at which the current is assumed to enter the cell, after having passed through the circuit outside is called the *negative terminal or pole.* The positive pole is generally indicated by a +, and the negative pole by a –. It is the plate of a voltaic cell which is connected to the negative pole that is dissolved in the electrolyte during the action of the cell, while around the opposite plate, gases tend to collect.

A very great variety of substances have been employed for the plates of voltaic cells. Some of the principal of these are zinc, copper, lead, carbon, platinum, iron and silver. Various liquids are used, both acids and alkalies. Among the most important of these are sulphuric and nitric acids, and caustic alkali.

We will now examine some of the principal voltaic cells in common use. Fig. 127, shows a simple form of single-fluid cell called the *bichromate of potash cell*. Here two plates, one of carbon and the other zinc, are dipped into a solution of bichromate of potash, to which a small quantity of sulphuric acid has been added. Binding posts a and

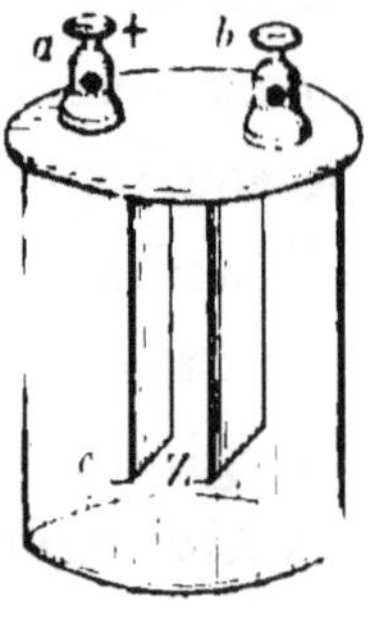

Fig. 127. Bichromate of potash cell.

b, connected to the plates, are provided for the ready attachment of the circuit wires.

Another form of single-fluid cell called a *Smee cell* is shown in Fig. 128. In this cell a plate of silver S, is placed between two zinc plates Z, Z. The electrolyte employed is sulphuric acid and water. This cell was formerly employed for *electroplating*, or covering a conducting surface with a metallic coating by electricity, but is now generally replaced for this purpose by dynamo-electric machines.

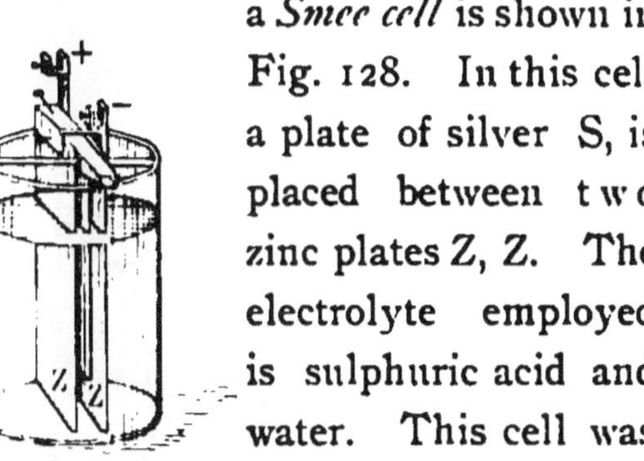

Fig. 128. Smee voltaic cell.

A very convenient form of single-fluid voltaic cell called the *Grenêt cell*, is shown in Fig. 129, where a plate of zinc Z, is placed between two plates of carbon K, K, and immersed in a solution of bichromate of potash in water, to which a quantity of sulphuric acid has been added. In order to prevent the zinc plate from being acted on, while the cell is out of action, the zinc plate is attached to the metal rod R, by means of which it can be raised out of the liquid.

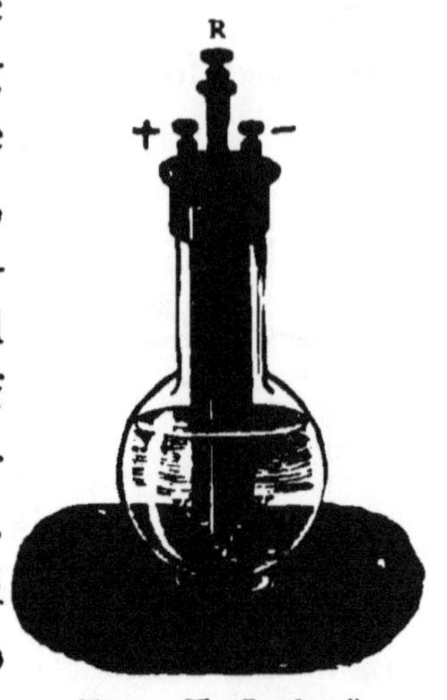

Fig. 129. The Grenêt cell

In double-fluid cells, in order to keep the two fluids separate, what is called a *porous cup* is employed. For example, in the voltaic cell shown in Fig. 130, called the *Partz cell*, a porous cup is used. The voltaic couple is formed of carbon and zinc. An exciting liquid, consisting of an aqueous solution of a salt called sulpho-chromic salt, is placed in the outer cup and a solution of common table salt, inside the porous cell. Sometimes, however, the two liquids are kept apart by means of their difference in density. For example, in the form of Partz cell shown in Fig. 131, the carbon

Fig. 130. The Partz cell, suitable for motor work.

plate C, is placed at the bottom of the cell, and the zinc plate is supported as shown near the top of the cell. The cell is partly filled with a solution of common table salt in water, when crystals of sulpho-chromic salt are added through the funnel F, and tube T. This salt falling to the bottom,

Fig. 131. Partz gravity voltaic cell

dissolves and forms a dense solution, which remains at the bottom of the cell. This cell produces an E. M. F. or pressure of about 2 volts.

In the ***blue-stone gravity cell***, shown in Fig. 132, a plate of copper C, is placed at the bottom of the jar, and a plate of zinc

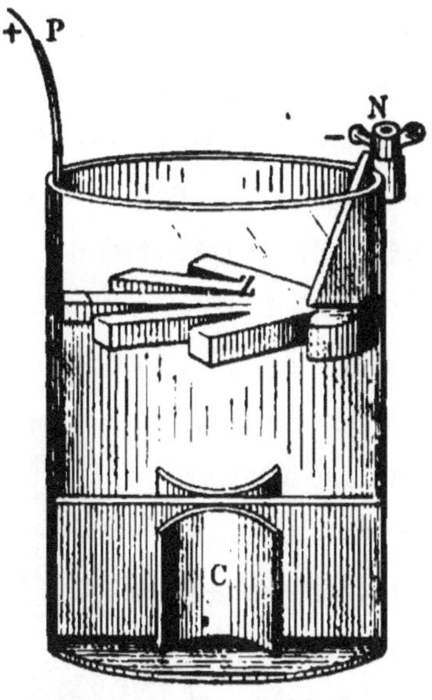

Fig. 132. Gravity blue-stone cell.

suspended near the top, in the manner shown. The copper plate is covered with crystals of blue-stone or copper sulphate, and a saturated solution of copper sulphate half fills the jar. A solution of sulphate of zinc is then poured on the top of the sulphate of copper solution, on which it floats because of its smaller density. The positive and negative terminals of the cell are shown at P and N, respectively. The blue-stone gravity cell is much used for telegraphic work, where it is desired to keep the cell on closed circuit for a considerable length of time. The blue-stone cell gives an E. M. F. of about 1 volt.

An excellent form of cell, suitable for use where current is not required for a very long time, and in which the cell has intervals of rest, is shown in Fig. 133. It is called the *Leclanché cell.* Its voltaic couple is formed of zinc and carbon. The

Fig. 133. Leclanché cell.

zinc is in the form of the rod Z, the carbon is placed inside the porous cell P, which contains a carbon plate surrounded by powdered peroxide of manganese. Both the porous cell and the outer jar contain a solution of sal-ammoniac in water. In this cell the oxide of manganese acts as a solid depolarizer. A Leclanché cell gives an E. M. F. of about one and a half volts. It is suitable for such purposes as do not require the current to be maintained for a long time.

The porous cell employed in the Leclanché cell is made of unglazed earthenware. Various attempts have been made to dispense with the porous cup in cells of the Leclanché type. In the cell shown in Fig. 134, called the *Gonda prism cell*, the necessity for the porous cup is dispensed with by holding or cementing the materials surrounding the carbon plate together by the admixture of some suitable resin, and then moulding them into the desired form under great pressure.

Another form of cell consists in a couple formed of zinc and carbon. In the form shown in Fig. 135, called the *Law cell*, an

Fig. 134. The Gonda cell.

extended surface is given to the carbon by making it of a number of separate carbon rods. The exciting fluid is a solution of sal-ammoniac in water.

Where a voltaic cell has to be carried

about, in order to avoid spilling the liquid, what are called *dry cells* are employed. The term dry cell is a misnomer. Such

Fig. 135. Law cell.

cells consist of various couples surrounded by a moist gelatinous substance containing an electrolyte. Two forms of dry cells are shown in Figs. 136 and 137.

Where it is necessary to obtain a greater voltage than a single voltaic cell is capable

Fig. 136. Dry cell.

Fig. 137. Dry cell.

of producing, a number of cells are connected in series to form what is called a *voltaic battery*. In this case the voltage or pressure will be equal to the sum of the voltages of the separate cells. Thus, in Fig. 138, three Leclanché cells are shown connected in series. Here the zinc terminal of cell No. 1, is connected with the carbon

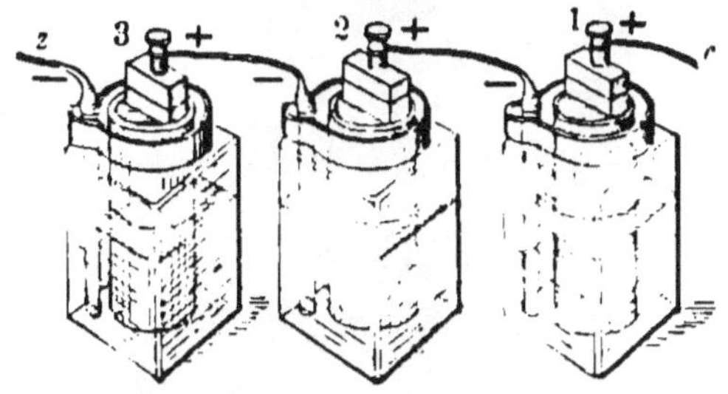

Fig. 138. Series-connected battery of three Leclanché cells.

terminal of No. 2; the zinc terminal of No. 2, is connected with the carbon terminal of No. 3; the free carbon terminal of cell No. 1, and the free zinc terminal of cell No. 3, form the terminals of the battery. In Fig. 139, a battery of 16 series-connected cells is shown. Here the cells are what are called *silver chloride cells*. This battery is suitable for medical use.

The word *battery* is sometimes applied to a single voltaic cell. This, however, is incorrect and should be avoided. In all cases the word battery should be limited

Fig. 139. Battery of 16 silver chloride cells.

to any combination of separate electric sources as will permit it to act as a single electric source.

Where it is desired to obtain a fairly

strong current for occasional use, a form of *plunge battery* is employed. Such a battery is shown in Fig. 140. It consists, as shown, of means whereby all the couples

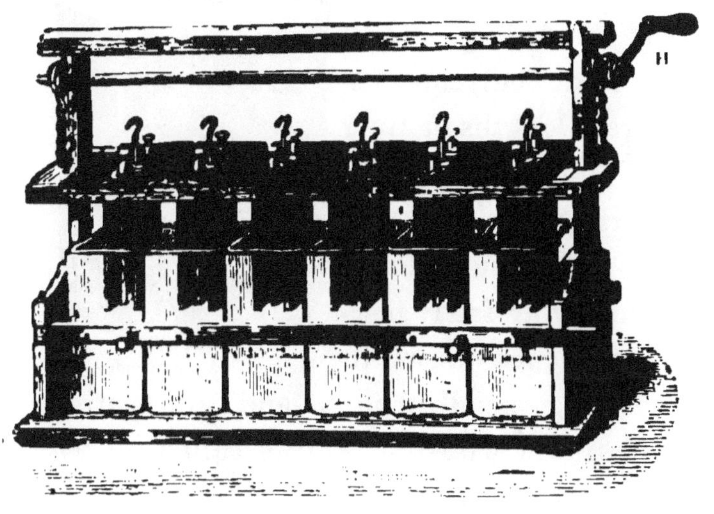

Fig. 140. Plunge battery.

may be raised from the liquid by turning the handle H.

In an excellent form of cell called the Edison-Lalande cell, the elements are formed of a plate of zinc and a plate of copper covered with copper oxide. The exciting liquid is a solution of caustic potash in water. This cell is capable of producing strong electric currents for great lengths of time without marked polarization, and is, therefore, suitable for supplying fairly powerful currents, as in driving small motors and in heating platinum wires or knives called *electric cauteries.* A form of this cell is shown in Fig. 141. A battery of three cells, connected in series, and intended for cautery work is shown in Fig. 142.

Fig. 141. Edison-Lalande cell.

Fig. 142. Battery of three series-connected Edison-Lalande cells for cautery work

CHAPTER XII.

THE ELECTRIC BELL AND HOW IT OPERATES.

THE piece of apparatus shown in Fig. 143, is probably well known to all our readers as a *magnet*. It consists of several bars of hardened steel firmly secured together by screws, and magnetized so as to produce a *north-seeking pole* at n, and a *south-seeking pole* at s. a, is a piece of soft iron, called the *keeper*. A magnet of hardened steel will retain its magnetism for an indefinite time, especially if its keeper a, be so kept on as to join, or connect, its two poles. Such

H

Fig. 143. Permanent magnet.

a magnet differs greatly from what is called an *electro-magnet.* Here, the magnetism can be turned on or off at will, by simply closing or opening an electric circuit, that is interrupting or completing it. Whenever an electric current flows through a conductor, that conductor becomes magnetic, and retains its magnetism as long as the current continues to flow, but loses its magnetism instantly, as soon as the current ceases to flow.

If a coil be wound around a bar of soft iron, as shown in Fig. 144, and the terminals of the coil be so connected with a voltaic cell, that the current flows in the direction indicated by the arrows, then, as soon as the current passes, the bar or *core*, as it is generally called, becomes magnetized, with a north-seeking pole at N, and a south-seeking pole at S. Such a magnet, consisting of a magnetizing coil provided with a soft iron core, so that the magnet can instantly acquire or lose its magnetism, is called an *electro-magnet*. If a bar of hardened steel be used instead of soft iron, it will retain its magnetism after the current has ceased to pass.

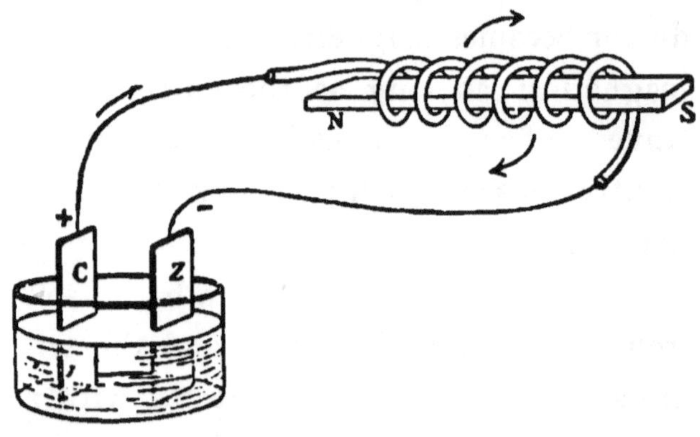

Fig. 144. Electro-magnet.

The fact that soft iron can acquire and lose its magnetism almost instantly, renders the electro-magnet of great value in a variety of signalling apparatus. A valuable form of signalling apparatus is called the *electric bell.* Before describing it we will look for a moment into the construction of an electro-magnet.

In any electro-magnet, such for example as the form of magnet shown in Fig. 144, the poles N and S, at the ends of the bar, have a strength or power of attraction which depends on the strength of the current that passes through the coils of the magnet, and also on the number of turns of wire in these coils. If, with the same number of turns and the same current strength, the iron core, instead of being straight, be bent, as shown in Fig. 145, the weight W, it can carry will be greatly increased.

It is generally found more convenient in practice to make the electro-magnet in separate pieces, as shown in Fig. 146. Here the core is formed of two straight rods a, a, connected by a *yoke piece* Y, of soft iron. The magnetizing coils C C, are placed on the core a a, in the manner shown.

The position of the N-seeking and S-

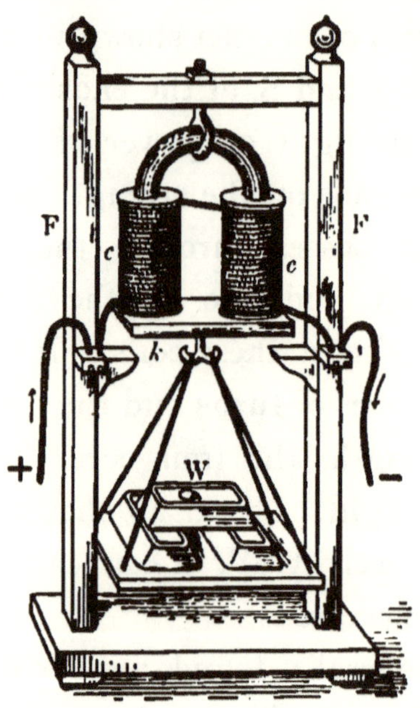

Fig. 145. Electro-magnet.

seeking poles of an electro-magnet depends both on the direction in which the magnetizing coils are wound, and the direction in which the current flows through the coils. Changing the direction of either the winding or the direction in which the current flows, changes the polarity. In electro-magnets, however, where soft iron keepers or armatures are used, a change in the polarity of the magnet makes no difference, the armature being attracted when the current begins to flow, and being no longer attracted as soon as the current ceases to flow, no matter whether the poles be N-seeking or S-seeking.

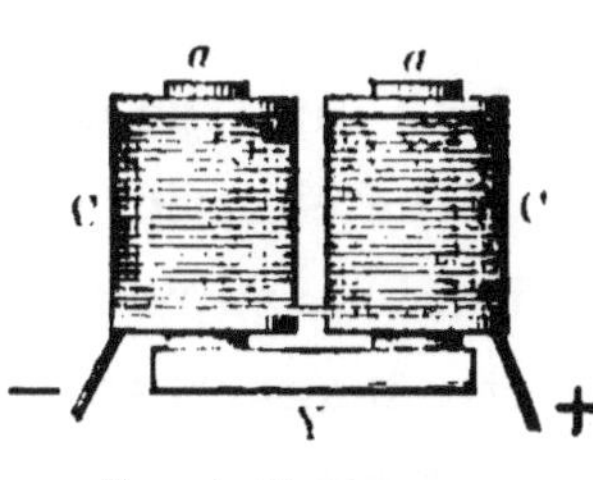

Fig. 146. Electro-magnet.

When a piece of soft iron is employed in connection with a magnet simply to aid it in retaining or keeping its magnetism, it is called a *keeper*, but when, in addition to this, it is so supported that it can be drawn or attracted towards an electro-magnet, as soon as the current passes, and drawn away from the magnet by the action of a spring or weight, as soon as the current ceases to pass, it is called an *armature*.

We are now ready to look into the construction and operation of the electric bell. Here, an armature is suitably supported near the poles of an electro-magnet, so as to be drawn towards the magnet when the current passes through the magnet coils, and to be drawn away from the magnet, by the action of a spring or weight, as soon as the current ceases to pass. A lever, provided at its end with a hammer or clapper, is so attached to the armature of the electro-magnet, that when the armature is drawn towards the magnet the clapper strikes the bell, and when the magnet ceases to attract its armature, the lever is drawn away from the bell, into a position in which it is ready to again strike the bell when attracted by the magnet.

Electric bells may be divided into two classes; viz., *continuous-striking* or *trembling bells*, and *single-stroke bells.* The first will continue striking when once set in action until the current is turned off. Such bells are suitable for calls and alarms generally. Single-stroke bells, as the name indicates, give only a single stroke when the current passes through the coils of their electromagnets. They are suitable for such special signalling as might be required in mines, railroad stations or other similar places.

Let us now examine the means whereby an electric bell can be caused to continue trembling or vibrating as long as the electric circuit is closed at the push-button connected with it. This is effected by means of what is called an *automatic contact-breaker*, the construction and operation of which will be understood from an inspection of Fig. 147. A vibrating spring C, placed in a vertical position, bears a piece of soft iron B, directly opposite the core of the electro-magnet A. A screw S,

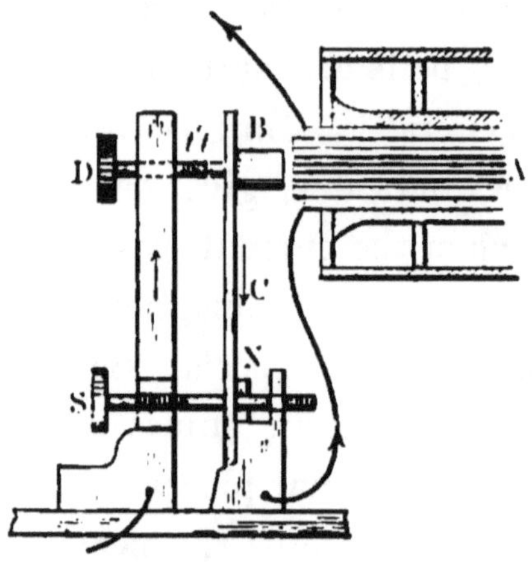

Fig. 147. Automatic contact breaker.

is provided for tightening the spring C, so as to vary its tension, as may be required. The spring C, is provided with a platinum contact piece t, directly opposite B, and on the side of the spring furthest from the magnet pole. A similar contact piece t′ is placed directly opposite the contact piece on C. If the circuit of a voltaic cell be connected with the electro-magnet through the contacts t,t′, then when the circuit is closed, say by the pressing of a push button, the current passes in the direction indicated by the arrows, and, passing through the magnetizing coils, the magnetism is turned on, and the magnet attracts the armature B, thus opening the circuit by breaking the contact between B and D. The magnet A, now loses its magnetism and the spring C, moves back under the action of its elasticity, thus again completing the circuit through the magnet coils, the magnet again attracting the armature, and again breaking the circuit. In this manner the armature will continue to move to-and-fro as long as the push button continues to be pressed.

In the continuous-ringing or vibrating bell, an automatic contact breaker is em-

Fig. 148. Continuous-ringing electric bell.

ployed similar to that shown in Fig. 147. A striking lever and hammer are attached to the armature A, which is pivoted at p.

Every time the current passes through the coils of the electro-magnet M, M, the armature is attracted, and the hammer

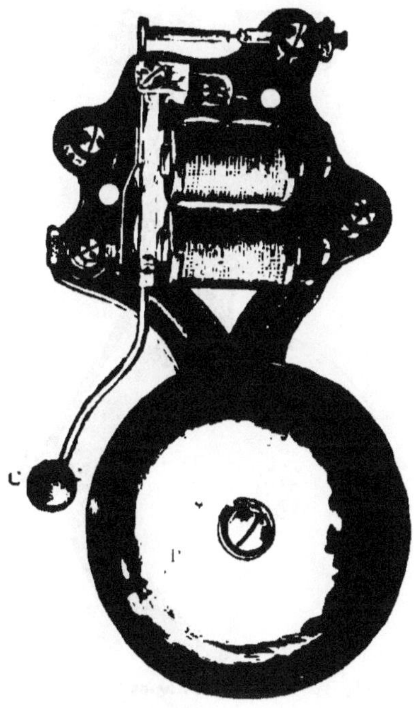

Fig. 149. Iron frame skeleton bell.

strikes the bell, and every time the current ceases to pass, it is drawn away by the action of the spring S. The contact piece is shown at C. Other forms of continuous ringing bells are shown in Figs. 149 and 150. In many cases continuous-ringing

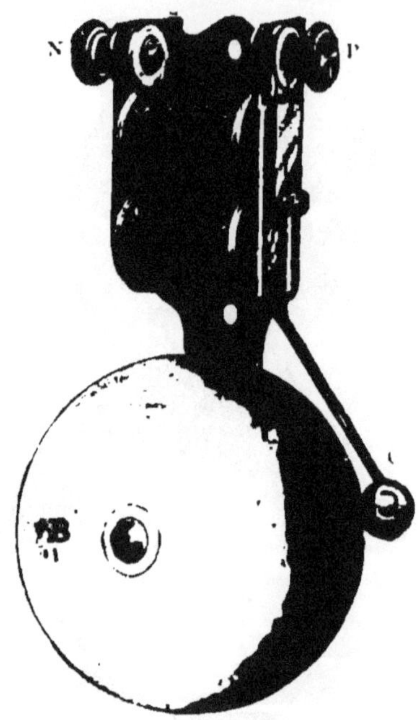

Fig. 150. Continuous-ringing bell.

bells can be changed into single-stroke bells by merely adjusting or cutting out of circuit the contact screw.

In the single-stroke bell the attraction

of the armature of the electro-magnet continues so long as the current passes through the circuit so that the bell will only give a single-stroke every time connection is made. As long as the current continues to pass, the armature remains in

Fig. 151. Single-stroke battery telephone bell.

contact with the cores of the electro-magnet. When the circuit is broken, the armature is drawn away from the magnet and is attracted when the current again passes. Such a bell is shown in Fig. 151. Single-stroke bells are used in cases where a number of different signals are to be sent. Another form of single-stroke bell is shown in Fig. 152.

In the operation of the electric bell the circuit is generally opened and closed by

Fig. 152. Single-stroke bell.

means of what is called a *push button* which is a device whereby a circuit may be closed by pushing a movable contact against the action of a spring until it comes in contact with a fixed piece. Various methods can be adopted for this purpose. That most frequently adopted is shown in Fig. 153, where a push button is shown with and without its cover. An

Fig. 154. Details of push button.

inspection of the figure will show that the mere pushing of B, will cause a movable spring piece p, which forms one terminal of the circuit, to come into contact with the fixed piece n, which forms the other contact.

It is a very simple matter to place an electric bell in circuit with a push button and a voltaic cell. The manner in which this is done is shown in Fig. 154. A wire W, is connected with one of the terminals of the cell, and one of the terminals of the push button. Another wire W′, is connected with the other terminal of the cell and one terminal of the bell. The other terminal of the bell is then connected with the other terminal of the push button by the wire W″. When the circuit is completed, by pressing the button P, the current flows in the direction indicated by the arrows. Since the voltage employed is small, generally being that of either a single cell or at the most of a few cells, in running a circuit, the wire, which should be insulated, may be supported directly

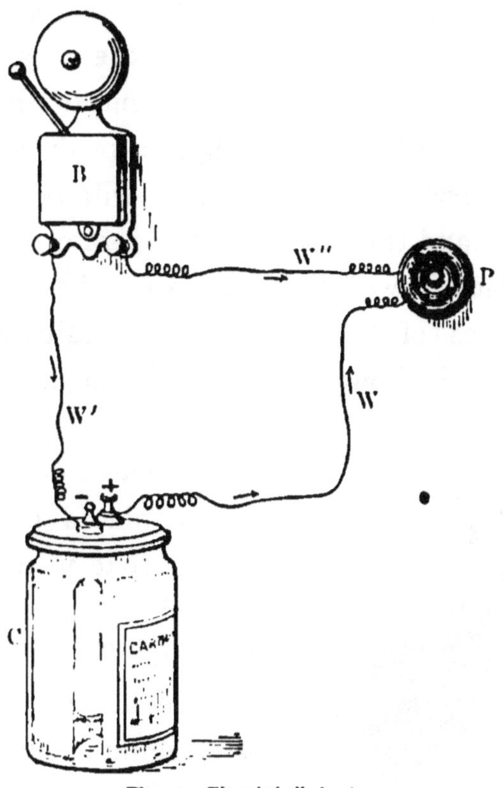

Fig. 154. Electric bell circuit.

on the walls of the room by means of staples. In connecting up the bell circuit it is necessary that the insulating material be carefully removed from the ends of the wires before connecting them to the terminals of the cell, bell and push button. Otherwise, the connection will be imperfect and the bell will not operate. Sometimes it will be found that the contact pieces of the bell require adjustment. This is, however, generally an easy matter.

Fig. 155 Floor push.

Sometimes a push button is placed on the floor so that it can be operated by means of the foot. This device is called a *floor push.* By placing the foot against P, Fig. 155, on the piece S, it completes the circuit by pressing against the two springs n and p.

Where it is desired to call an attendant quietly without the noise of an electric bell, a device called a *buzzer* is employed. This consists of an automatic electro-magnetic contact-breaker, the to-and-fro motions of whose armature are caused to produce a low musical note not unlike the buzzing of an insect. Hence the name buzzer. A form of buzzer is shown in Fig. 156.

Where it is desired to obtain a loud stroke of the bell without the use of a strong electric current, a device called an *electro-mechanical gong* may be used. Such a gong is shown in Fig. 157. It consists of means whereby a comparatively feeble electric current operating an electro-magnet is used to throw into action any simple mechanical device for striking a bell by the force of a spring, or by the falling of a weight.

Where it is desired to ring electric bells

without the use of a voltaic battery, the electricity can be generated as required by the use of a *magneto-generator*; i. e., a form of dynamo in which the field magnets

COVER OFF.

COVER ON

Fig. 156. Electro-magnetic pocket buzzer.

are permanent magnets. A bell so operated is called a *magneto bell.* Magnet bells are in common use for telephone work. When it is desired to ring a distant bell,

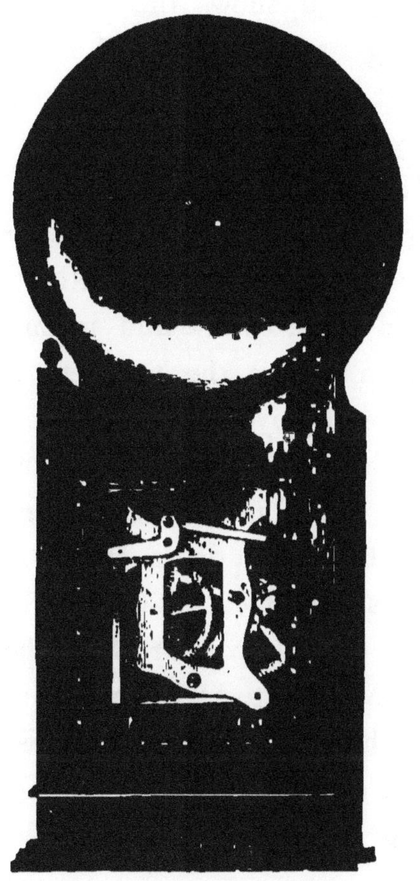

Fig. 157. Electro-mechanical gong.

the handle H, is rapidly rotated, thus rotating the armature which revolves between the magnet poles.

Fig. 159, shows the interior of a magneto generator suitable for ringing bells or for electrical testing purposes. Where very long or high-resistance circuits are to be tested, it is necessary to employ a high voltage, such as would require perhaps a hundred voltaic cells. In order to avoid the expense and inconvenience of such a battery, the magneto testing bell can be employed.

When an electric bell is placed in positions where an attendant cannot remain near the bell, a device called an *extension bell* is employed. This bell is located in some distant room or place where the person is likely to be when not in the office.

Where a voltaic battery is employed, since electric bells are only occasionally rung, the battery will have long periods

Fig. 158. Magneto signal bells.

Fig. 159. Magneto testing bell.

of rest. Any good form of battery, that will depolarize on open circuit if a sufficiently long time be given it, will therefore serve. The Leclanche battery, in some of its many forms, is excellent for this purpose.

Fig. 160. Extension bell.

We will now examine how a number of electric bells can be placed in the same circuit. Fig. 161, shows how a single push button P, is placed in the circuit of

Fig. 161. A single bell operated, by a single push.

a single bell B, and a voltaic cell E. When it is desired to operate this bell by two different push buttons, the circuit connections are as shown in Fig. 162, where two push buttons P′, and P, are connected in multiple to the circuit. Here the bell will be rung when either push is operated.

Sometimes it is necessary to operate two

Fig. 162. A single bell operated by two pushes.

different bells by a single push button. In this case the connections are as shown in Fig. 163. Here the two bells are connected as shown to the circuit in multiple.

Fig. 163. Two separate bells operated by a single push.

If it is desired to operate two different bells from two different buttons, the arrangement of circuits shown in Fig. 164, may be adopted. Here, as will be seen, the closing of the circuit of P, will operate the bell B, while the closing of the circuit of P′, will operate the bell B′.

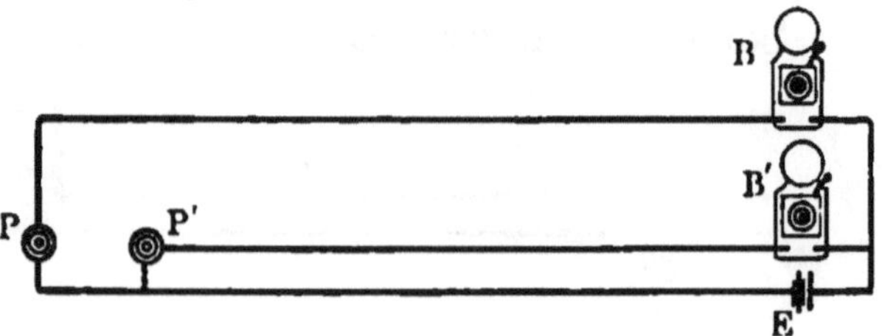

Fig. 164. Two separate bells operated by two separate pushes.

It is generally advantageous, on the ringing of the bell at the distant end of a line, for the person called to be able to

Fig. 165. Simple-button three-line return-call. One battery.

ring back or send a return call, or to be able to send the call from the other end of the line. In such a case the arrangement of circuits shown in Fig. 165, may be employed. In Fig. 365 three separate line wires or conductors are necessary with only a single battery E. The closing of the push button P, will ring the bell B, while the closing of P′, will ring B′.

By using the gas pipes, or the ground

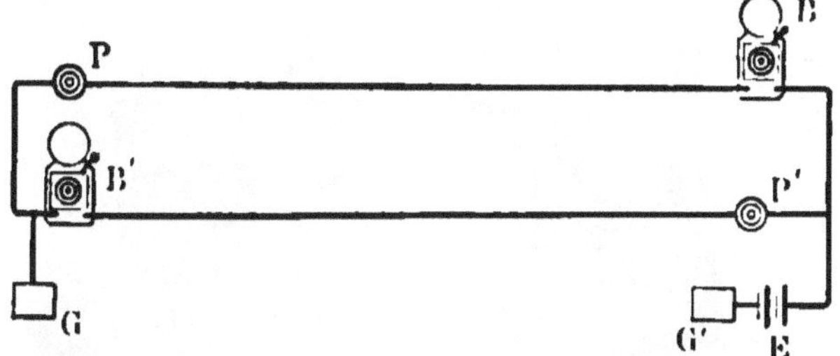

Fig. 166. Simple button two-line, ground-return call. One battery.

for a return wire, a return-call circuit can be obtained with 2 wires. This circuit is shown in Fig. 166, one end of the line wire or conductor being connected to the ground plate G, or to the gas pipe, and one end of the voltaic battery being similarly connected to a ground plate G′.

CHAPTER XIII.

The Electric Telegraph and How it Operates.

The electro-magnetic telegraphic receiving instrument, as invented by Morse, con-

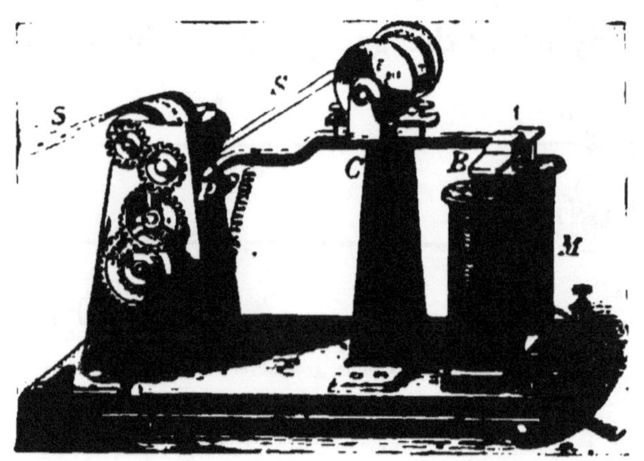

Fig. 167. Early form of Morse receiving instrument.

sisted, as shown in Fig. 167, of an electro-magnet M, provided with an armature B, to which is attached a lever pivoted at C. The lever is provided at its free end with a metallic point p. When an electric current passes through the coils of the electromagnet, the downward movement of the armature presses the point p, against a band of paper S S, kept moving under the point by clock-work. If the electric current passes through the coils of M, but for a moment, a dot or indentation will be made on the paper; but, if it continues to pass for a longer time, a line or dash will be made on it. By suitably interrupting the current at the sending end of the line, characters representing the letters of the alphabet are thus recorded in dots and dashes on the moving strip of paper.

In the *Morse telegraphic code* or *alphabet*, the letters and numerals are obtained by various combinations of a dot, a dash, and an interval or space. Fig. 168, represents the Morse alphabet as employed in the United States. It will be seen that the letter a, is represented by a dot, an interval and a dash ; s, is represented by three dots ; the numeral 6 is represented

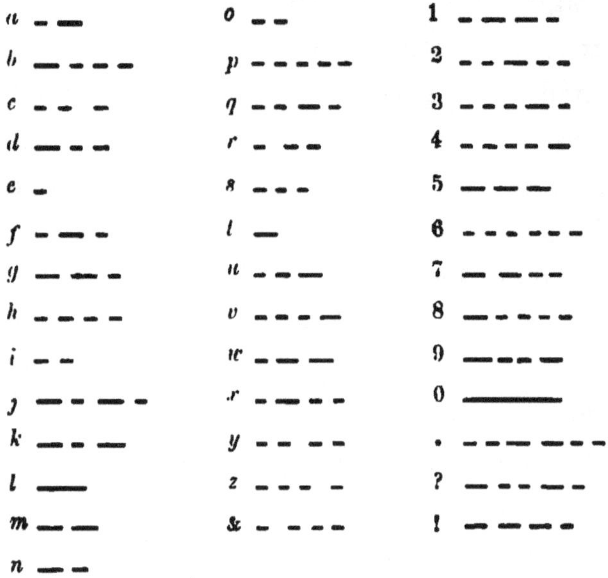

a	. -	*o*	. .	1	. - - .
b	- . . .	*p*	. . . . .	2	. . - . .
c	. . .	*q*	. . - .	3	. . . - .
d	- . .	*r*	. . .	4	. . . . -
e	.	*s*	. . .	5	- - -
f	. - .	*t*	-	6	. . . . . .
g	- - .	*u*	. . -	7	- - . .
h	. . . .	*v*	. . . -	8	- . . . .
i	. .	*w*	. - -	9	- . . -
j	- . - .	*x*	. - . .	0	——
k	- . -	*y*	. . . .	.	. . - - . .
l	——	*z*	. . . .	?	- . . - .
m	- -	&	. . . .	!	- - - .
n	- .				

Fig. 168. American Morse code.

by six dots ; r, by a dot and an interval of double length, and two dots.

In practice, it was found that an operator soon learned to read the message received by the sounds made as the lever moved to-and-fro under the alternate influence of the attraction of the electro-magnet, and the action of the spring provided to draw back the armature on the cessation of the electric current. The operator did not have to read the message from the record left on the paper. The registering apparatus was, therefore, removed and the instrument was converted into what is called a *telegraphic sounder.* A Morse sounder is shown in Fig. 169. To the armature A, of the electro-magnet M, is attached a striking lever L. When the current passes through the coils of the magnet M, the armature is attracted until the end of the stop screw S, strikes against the metal stand O. As soon as the electric current ceases to pass, the armature is drawn away from the magnet by the action of the spring Q, until it strikes against the

Fig. 169. Telegraphic sounder.

end of the stop screw S'. These two sounds differ from each other and are soon recognized by the operator as Morse characters when they follow one another at the proper intervals. The tension of the spring Q, and, consequently, the intensity of the blow given to the lever L, on the movements of the armature A, can be regulated by the screw R.

Let us now examine how the intermittent currents are sent over a telegraphic line so that Morse characters are recorded on the paper strip, or produce the characteristic sounds in the Morse sounder. The apparatus for this purpose is called a *telegraphic key*. A form of *telegraphic key* designed for attachment to a table is shown in Fig. 170. A metal base B B, is firmly screwed to the top of a table by means of the screws O, O,' that pass through the table. The ends of the line wires are connected to binding screws

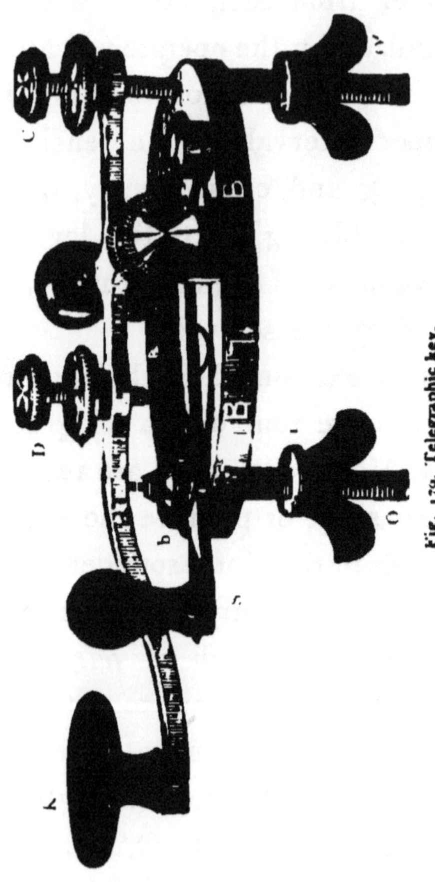

Fig. 170. Telegraphic key.

O, O′. O, which is carefully insulated from the base, terminates in the platinum contact piece b; while O′, is in good electric contact with the base. The key K, has its lever in contact with the base, and is, therefore, in contact with O′. The key lever is provided with a platinum contact piece at a, immediately opposite b. The circuit can be closed either by depressing the key K, until a and b, are in contact; or by closing the switch S.

The regulating screw D, is provided for varying the tension of the spring R, which causes the key to move back when it is released after being depressed. The screw C, regulates the distance through which the key can be moved before the contact piece strikes against its stop.

In Fig. 171, is shown a complete Morse telegraphic set, with a voltaic cell, key and sounder connected in the circuit. Here the cell is represented as quite near

Fig. 171. Morse telegraph outfit for amateurs.

the instrument, but of course is generally at a considerable distance from it—often of several miles.

A comparatively strong current is necessary to be passed through the coils of the telegraphic sounder in order to make distinctly audible signals, or to mark the characters on the paper slip in the registering apparatus. Consequently, when a telegraphic line is of considerable length, the current which enters the receiving apparatus may not be strong enough to operate it properly; in order to obviate this difficulty, an instrument called a *telegraphic relay* is employed in place of the sounder.

The telegraphic relay consists of an electro-magnet, the attraction of whose armature is made to open or close the circuit of a battery called a *local battery* in which the telegraphic sounder is placed. The operation will be easily understood from an inspection of Fig. 172, where R, is the relay placed in the line circuit L L. The attraction of its arma-

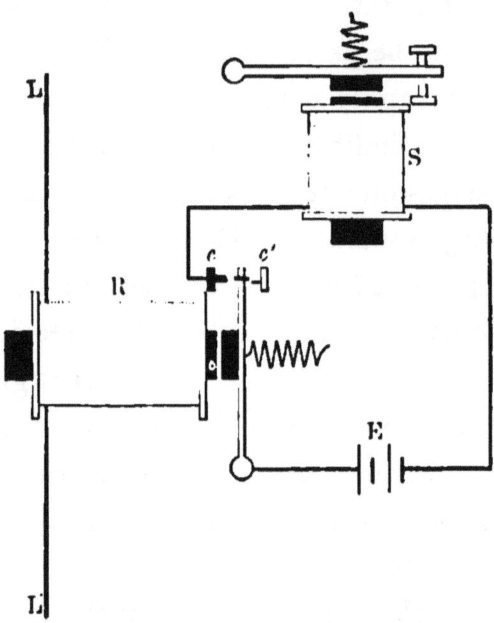

Fig. 172. Connections of relay and sounder.

ture closes the contacts at c, c,′ thus completing the circuit of the local battery through the sounder S.

A form of telegraphic relay is shown in

Fig. 173. The terminals of the line wires and of the sounder circuit are shown at 1, 1, and 2, 2, respectively. The armature A, through the lever *l*, closes the contacts a and b, connected with the circuit of the local battery and sounder.

Fig. 173. Telegraphic relay.

Sometimes, in fairly quiet offices, the sounds produced by the relay are sufficiently loud to permit the signals to be read directly, provided the instrument is surrounded with a resonant box. Such a relay is called a *sounding relay* but is, in reality, only a sensitive form of sounder. A sounding relay together with its key are shown in Fig. 174.

Fig. 174. Bunnell's box sounding relay and key.

The arrangement of a simple telegraphic line with ground return is shown in Fig. 175. Here there is placed at each end of

Fig. 175. Simple telegraphic circuit with ground return.

the line a telegraphic key and a sounder. The battery is represented as being placed at one end of the line only. On long lines it is usual to employ a battery at each end of the line. A and B, are the two stations, m and m, the two sounders; k and k, the keys, and E, the voltaic battery. When A, wishes to send a message to B, he opens his switch, and sends the makes and breaks in the order necessary for producing definite Morse characters. The currents passing through the line immediately cause similar movements in the armature of the sounder at B. As soon as A, has finished sending, he closes his switch, so that B, can answer him if he so desires.

CHAPTER XIV.

HOW THE DYNAMO OPERATES.

ALTHOUGH the exact operation of a dynamo-electric machine is quite a complicated matter, yet the general manner in which it produces electric pressure, by the rotation of its armature coils is very simple. Suppose a coil of insulated wire C, Fig. 176, has its terminals a and b, connected to the *galvanometer* G, which is a device for indicating the presence of an electric current in its coils. Then if one end or pole, say the N-pole of the magnet M, be thrust into the coil, an E. M. F. or pressure will be temporarily produced in it, and a current will flow through the galvanometer in a certain direction as will be indicated by the movement of the needle of the galvanometer. When the magnet is drawn out of the coil, an E. M. F. and current are also produced, but in the opposite direction. If the other

Fig. 176. Magnetically induced current.

end of the magnet; i. e., its other pole or its S, pole be thrust into or removed from the coil, an E. M. F will also be set up; but the E. M. F's and currents produced by an N, pole are always in the opposite direction to those produced by an S, pole. E. M. F's produced in this way are said to be produced by *magnetic induction.*

The cause of the E. M. F's is as follows: —There always surrounds a magnet, a peculiar streaming or condition in the ether called *magnetism* or *magnetic flux*. Whenever magnetic flux is caused to pass through a conducting loop of wire, an E. M. F. is set up in the loop. Now when a magnetic pole is thrust into a coil, the flux accompanying the magnet enters the coil and sets up an E. M. F. in it, and when the magnet is withdrawn, the loop is emptied of its flux, and an opposite E. M. F. is produced, the E. M. F. produced by emptying a loop of its flux being always oppositely directed to that produced by filling it with flux. The same effects will be produced if the magnet remains fixed and the coil is moved towards and from the magnet. It is important to observe, however, that if the magnet and the coil are stationary, relatively to each other, no E. M. F. is produced in the coil, no matter how much magnetic flux may be passing through the loops. It is only while the quantity of magnetism or magnetic flux passing through the loops is being changed that E. M. F. is developed.

Magnetic flux is assumed to come out of a magnet at its north-seeking pole, and to reënter it at its south-seeking pole. The direction of the flux streams which surround a magnet can be shown by sprinkling iron filings on a glass plate placed over the magnet, and then gently tapping the plate so as to aid the filings in arranging themselves in the directions of the flux streams. Such a grouping is shown in Fig. 177, between the N. and S. poles of two bar magnets. Here the flux is seen to follow curved paths between the opposite magnet poles.

If instead of inserting a magnet and its accompanying magnetic flux into a loop, the loop be moved into and out of the

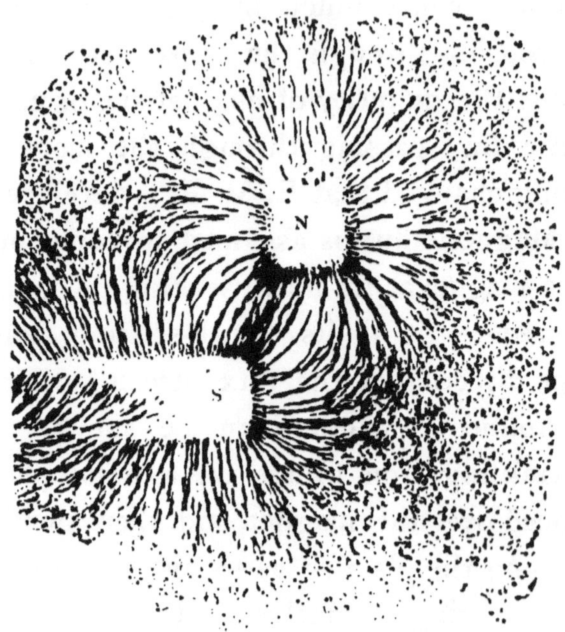

Fig. 177. Magnetic flux between N. and S. poles.

magnetic flux surrounding the magnet, E. M. F's will also be induced in the loop. If, for example, a loop of wire be rotated between two opposite magnetic poles,

E. M. F's will be set up in the loop, the device forming a simple *dynamo.* A dynamo consists essentially of a magnet for producing magnetic flux, and of coils of wire capable of being rotated in this flux.

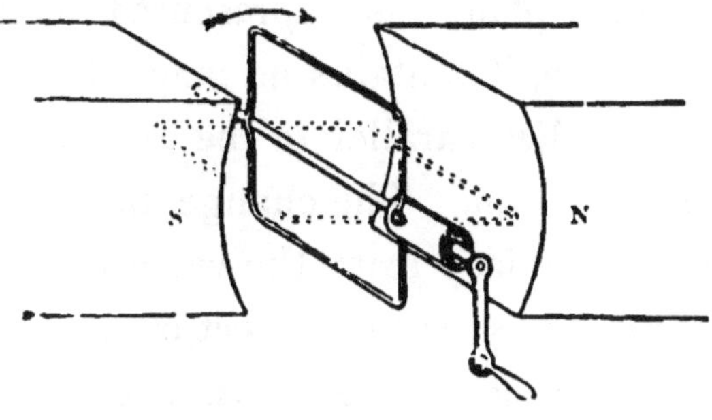

Fig. 178. Simple dynamo.

Fig. 178, shows a pair of iron magnet poles N. and S. forming part of a permanent magnet, or of an electromagnet. In the air gap between these poles, a powerful magnetic flux passes from the north to the south pole. If a loop of wire be supported about an axis between the poles as shown, and be rotated in the magnetic flux in the direction of the arrow by the handle, E. M. F's. will be set up in the loop. It will be evident that when the loop is in the vertical position shown by the full lines, it is filled with magnetic flux. When, however, the loop is in the horizontal plane, as represented by the dotted lines, it contains no magnetic flux, because it lies parallel to the direction of the flux stream. The change in the position of the loop from the vertical to the horizontal has had the effect of removing all the magnetic flux contained in the loop, and this will cause an E. M. F. to be induced in the loop. Continuing the rotation of the loop, will cause it to be alternately filled and emptied with magnetic flux and so cause E. M. F's. to be induced in the loop. These E. M. F's would be unable to send any useful current outside the machine if the loop were closed upon itself. They would only be able to send currents around the loop. If, however, the loop, instead of being closed, be opened at some point, say near the handle, and the two ends of the loop be connected with an external circuit, the E. M. F's. induced in the loop will be able to send electric currents through the external circuit so provided. In order to maintain a continuous electric action, between an external circuit and the revolving loop, it is necessary to employ brushes resting upon a contact cylinder attached to the revolving loop. Such a contact cylinder is shown in Fig. 179. When used as shown, it not only enables connection to be per-

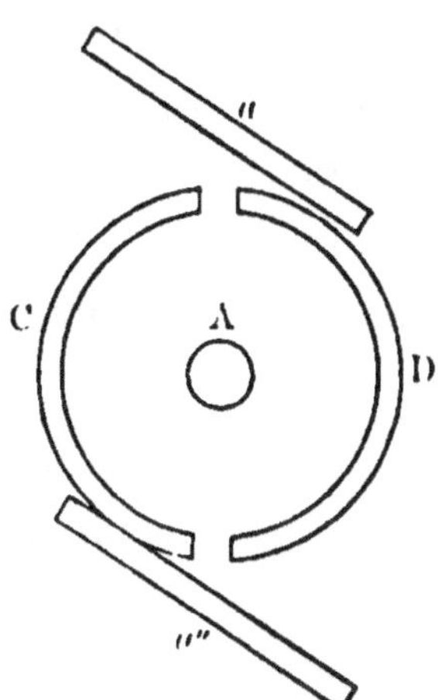

Fig. 179. Simple two-part commutator.

manently established between the wire in the revolving loop and the wire in the external circuit, but it also provides for the delivering of the current in the same direction to the external circuit, that is, to deliver direct or continuous currents and for this reason it is called a *commutator.* A simple two-part commutator, suitable for use with a single revolving loop, is shown in Fig. 179. Here the ends of the loop are connected to the segments C and D, and the brushes a, a″, rest upon opposite points of the diameter. These brushes are connected with the external circuit, and through them the current passes into and out of the armature.

In actual dynamos more than a single loop is used in the revolving armature. The E. M. F.'s in these loops are all added, so that the total E. M. F. or *voltage* of the machine is increased by increasing the number of loops which are revolved through the magnetic flux.

Fig. 180, shows a form of small dynamo

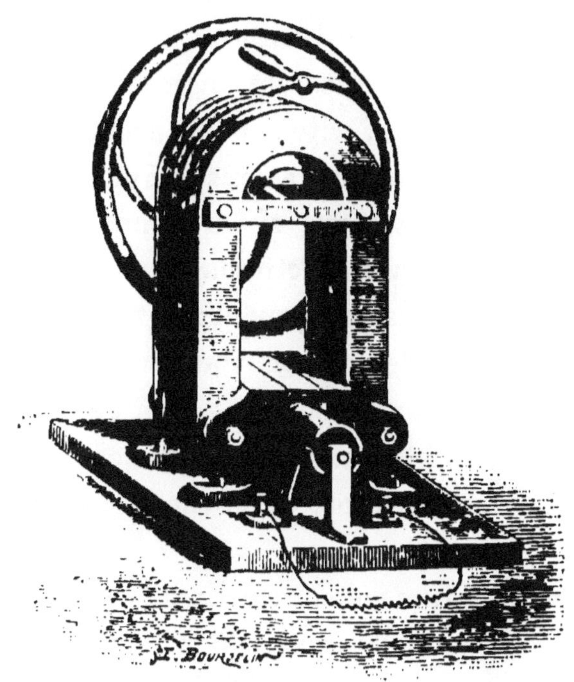

Fig. 180. Magneto-electric machine.

with a two-part commutator. Here, the field magnets are formed of bars of hardened steel.

The E. M. F. produced by a dynamo can be increased by increasing the speed of rotation with which the armature coils are emptied and filled with magnetic flux. It can also be increased by increasing the number of coils of wire on the armature. Generally, the armature coils of dynamos, consist of many turns of wire, and their field magnets are electromagnets. The number of pieces or segments in the commutator will depend on the number of magnet poles, the number of coils on the armature, and the manner in which the armature coils are placed on the armature core.

The armature coils are generally placed on a core of soft iron, which increases the quantity of magnetic flux passing through the coils. This increase is due to the iron permitting the flux to pass through it more readily than would any other substance, and also because the presence of iron increases the quantity of such flux. In order to prevent wasteful currents from being set up in the iron of

Fig. 181. Dynamo-Electric-machine.

the armature core, it is built up of a number of thin plates. Such a core is said to be *laminated*.

Fig. 181, shows a common form of

dynamo in which the field magnets have two poles N and S, between which revolves the armature A. The commutator C, is supported on the same shaft as the armature, and consists of a number of bars of hard copper, insulated from each other by strips of mica, but connected with the successive terminals of the armature coils all the way round. The brushes B, B, are of copper strips, though for large machines, brass gauze, or carbon brushes are frequently employed. These brushes rest upon the surface of the commutator, under the pressure of springs attached to the brush-holders. The pairs of brushes are placed at opposite ends of the diameter as shown in Figs. 179 and 180. One of these pairs of brushes, say the upper pair, is positive, while the lower pair is negative. Instead of employing permanent magnets of hard stee' the magnets are electromagnets, which are kept permanently magnetized by the passage of a continuous current. The pulley P, is employed to drive the armature at the proper speed.

A small dynamo, weighing say 100 pounds, can readily be made to give a pressure of 100 volts at its brushes. The current, however, that it can produce, is necessarily limited by the conducting capacity of the wire wound on its armature; since, if too strong a current be passed through this wire, it would be unduly heated. This limitation in the output of a dynamo corresponds to the limitation in the water current delivered by a pump; for, while a small pump could readily be made to produce the same pressure as a larger pump, yet the flow of water it could deliver, would be necessarily limited by that which could pass through the pump itself. Consequently, a dynamo which will deliver a powerful current as well as a high pressure, is necessarily a machine with large wire wound upon its armature, and, therefore, having a large mass of iron in its structure; i. e., a heavy machine, so that the dynamos in central stations often attain very considerable proportions.

CHAPTER XV.

HOW THE ELECTRIC MOTOR OPERATES.

If instead of expending mechanical energy for rotating the armature of any dynamo and thus setting up E. M. F.'s, and, consequently, electric currents therein, we expend electric energy by suitably passing electricity through the dynamo, its armature will rotate and become a source of mechanical power. If, therefore, we connect a dynamo at each end of a conducting line or circuit and drive one of them by mechanical power, the electric currents produced will flow through the line and cause the other dynamo to turn as a motor, thus developing mechanical power. In this way we can transmit mechanical power from one end of the line to the other, even over distances of several hundred miles.

The cause of the rotation of the electric motor is to be found in the magnetic pull which is exerted between the armature and the magnet between whose poles it rotates. It is a well-known fact that a suitably supported magnetic needle will be attracted to the pole of a magnet near which it is brought. It will, however, come to rest in the nearest position it can get to this magnetic pole. A continuous rotary motion, therefore, could not be produced in this way by a permanent magnet unless some means were known by which the magnetism of permanent magnets could be rapidly reversed, and there are no known means by which this can be done. But we can easily change the polarity of an electro-magnet by changing the direction in which the current passes through its magnetizing coils. Therefore, if any arrangement be devised, by which the polarity of the magnets is reversed at suitable intervals, it will be possible to obtain a continuous rotary

Fig. 181. Froment motor.

motion by magnetism, and this in fact constitutes the electro-magnetic motor.

An early form of *electro-magnetic motor* is shown in Fig. 181. Here the armature consists of a wheel supported on a horizontal axis and bearing on its periphery bars of soft iron a, a, a, etc. On the passage of an electric current from the battery B, through the motor, one set of magnet poles m, m, attracts the armatures a, a, but as soon as the armatures come opposite the magnet poles, the current is cut off from these magnets by means of a commutating device and sent into another set of magnets, which in their turn are cut out of circuit and so on in succession, so that the armature is kept in a continuous rotation. A motor of this type would, however, compare but poorly with the improved motors that are in such common use to-day, which not only employ electro-magnets both in their field magnets and armatures, but also have all of their magnet poles acting at the same time.

It may be well to remember that many of the electric devices we have already described are in point of fact electric motors, and their operation necessitates a transmission of power. For example, the electro-magnetic bell is in reality a motor. We produce an electric current, say by a voltaic cell situated at one end of a line, and cause this current to do work in moving the armature of an electro-magnet situated at the other end of the line. So too, a Morse telegraph receiving apparatus or relay, is a motor, and its operation similarly necessitates an electric transmission of power.

The power which an electric motor exerts depends, of course, on the speed with which it runs and the pull it exerts at its pulley. This pull depends on the amount of magnetic flux which passes through the armature, on the strength of the current supplied to the motor, and on the number of coils of wire placed on the armature.

The electric motor, as made to-day, is a very efficient piece of apparatus. Efficiencies as high as 90 per cent. can readily be obtained in motors of fairly large sizes. That is to say, a motor which receives one hundred horse power in electric power could readily deliver ninety horse power in mechanical power at its belt or pulley. This fact, together with the ease with which electric motors can be installed and operated, have caused them to be very generally introduced. Their use for propelling trolley cars is well known. Motors for this purpose are generally constructed so as to be operated by an electric pressure of about 500 volts. The current required for their operation is taken from the trolley wire or conductor by means of the trolley wheel which runs along under such conductor. The motors are placed underneath the car; and, in order to protect them from dust and mud, they are generally enclosed in water-tight, dust-proof boxes. A form of electric motor, employed for street-car work,

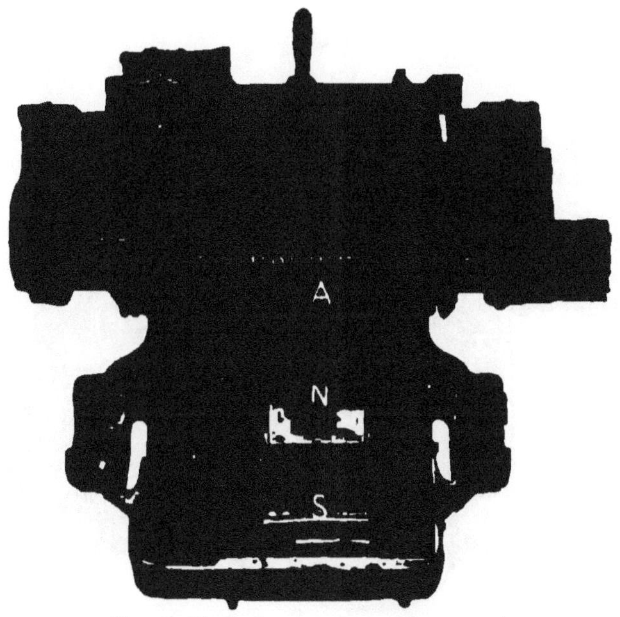

Fig. 182. Westinghouse street-car motor, opened.

is shown in Fig. 182, with the cover or lid open to show the interior. The field magnet poles are shown at N and S. A, is the armature. The brushes rest on the commutator. The armature is coupled to the car axle through gear wheels.

Figs. 183 and 184, show a modern form of electric motor assembled and disassembled.

Fig. 183. Electric motor.

Similar parts are marked with similar letters of reference. It is of the *bi-polar* type; i. e., the magnets have two poles only. The magnetizing coils are

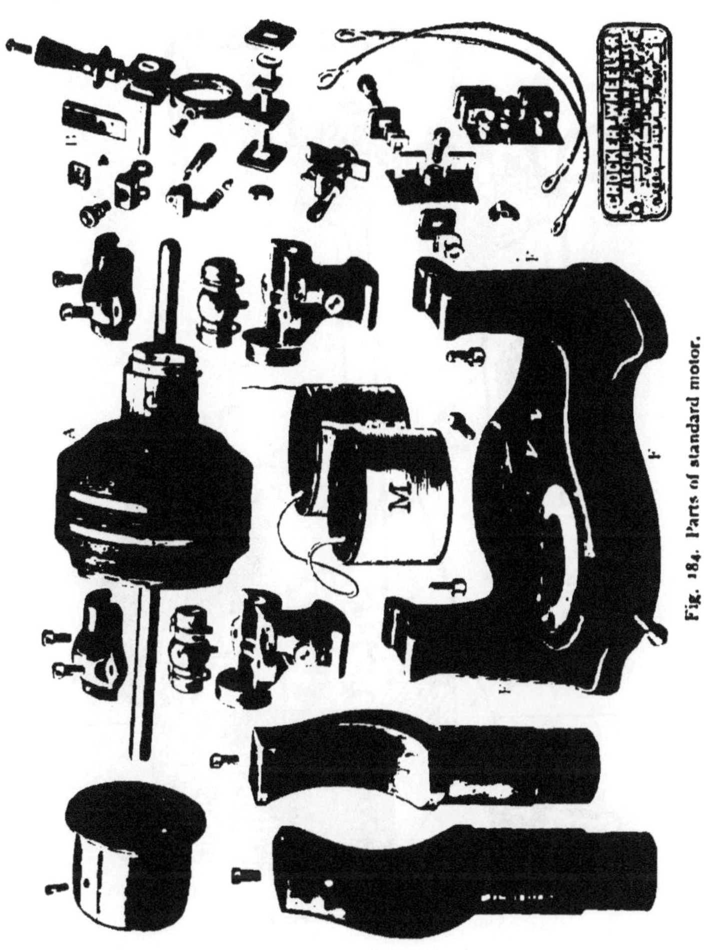

Fig. 184. Parts of standard motor.

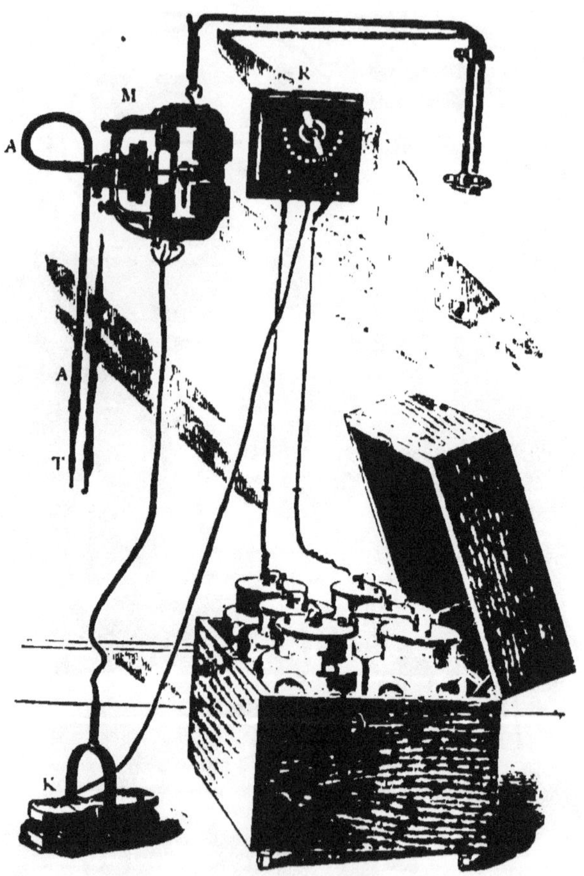

Fig. 185. Edison dental motor outfit.

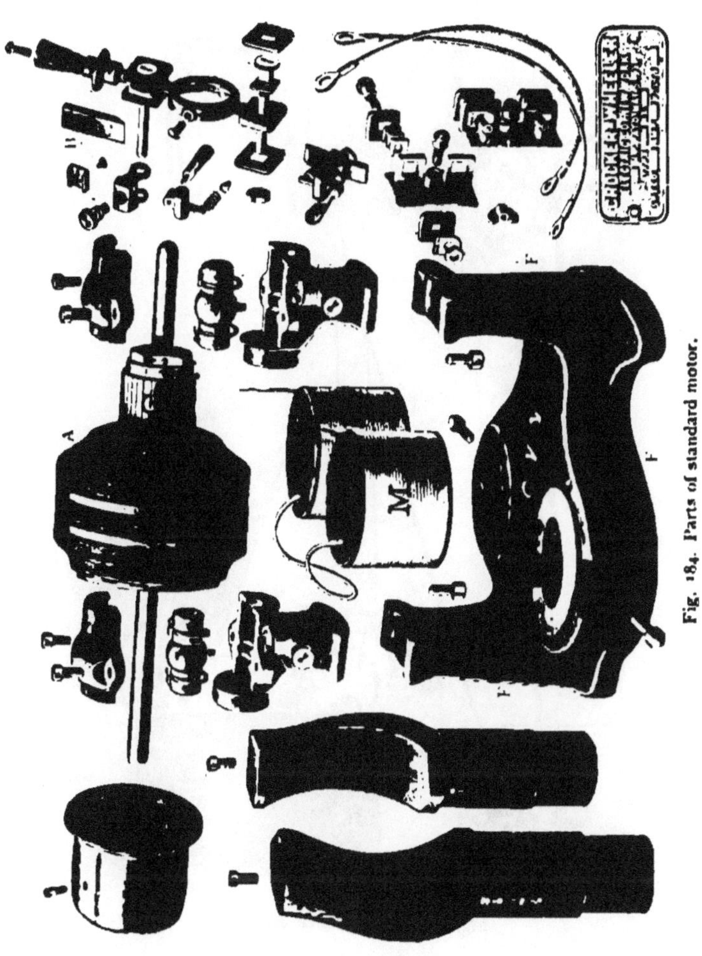

Fig. 184. Parts of standard motor.

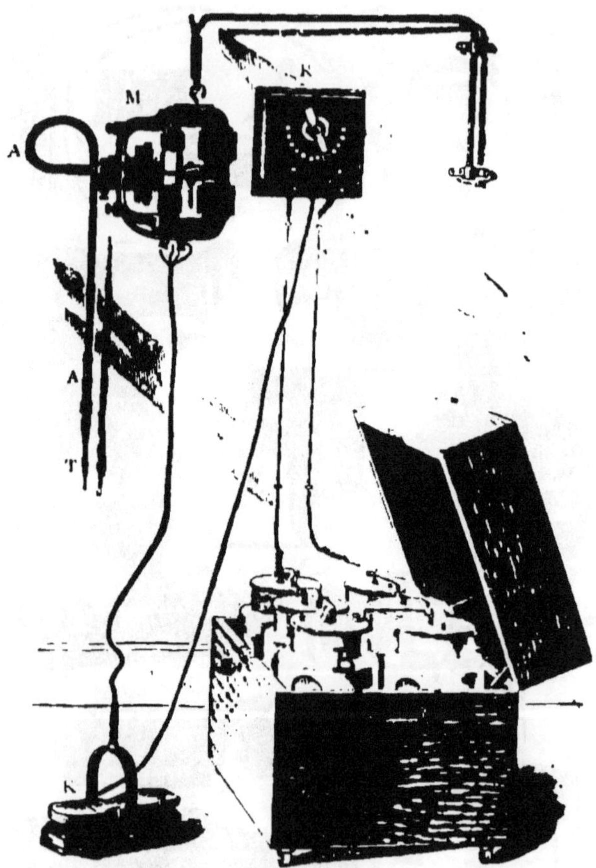

Fig. 185. Edison dental motor outfit.

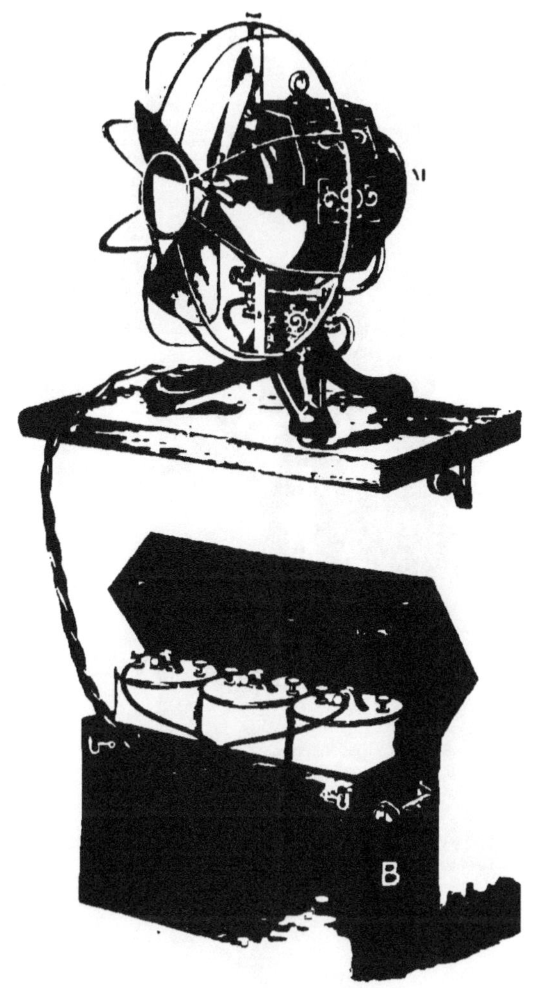

Fig. 186. Fan motor and battery.

Fig. 157. Electric motor directly attached to lathe.

shown at M, M. The commutator is shown at C, with the two sets of collecting brushes B, B′. The pole-pieces of the field magnets are shown at N and S, and the field frame at F, F, F.

In Fig. 185, is shown the application of a voltaic battery B, of six Edison-Lalande series-connected cells, to a dental motor outfit. The motor is shown at M, with an instrument T, attached to it through a flexible arm A, A. A *starting switch* and rheostat are shown at R. K, is a *foot-switch* provided for ease of starting the motor. Fig. 186, shows a battery B, of three series-connected cells intended for use with a fan motor M.

A great advantage of the electric motor consists in the fact that it can be placed directly on the shaft of the machine to be driven, thus dispensing with the use of belting or gearing. Fig. 187, shows an electric motor directly connected to a lathe.

CHAPTER XVI.

THE TELEPHONE AND HOW IT OPERATES.

PERHAPS no greater or more useful electric invention has ever before been produced than the *speaking telephone.* Like other great inventions, it was the product of many minds, beginning with the instruments produced by Reiss in 1861, continuing with the instrument of Bell in 1876, and the improvement is still going on. The telephone has done more to change business methods than has any other invention. By its use one is able to pass through space almost in the twinkling of an eye; to enter, in effect, the office of any desired correspondent, and, after a more or less prolonged conversation, to immediately re-enter his own office.

Our limited space will prevent us from entering into a detailed description of this wonderful instrument. It will suffice, however, to call attention to the fact that the telephone consists, at one end of a line, of what is practically a dynamo-electric machine driven by the voice of the speaker, and, at the other end of the line, of a small motor, by means of which the voice-produced electric currents reproduce the voice of the speaker. The motor and the dynamo; or, as they are called, the *telephone transmitter* and *receiver*, are placed at the ends of a conducting line and may be hundreds of miles apart.

When we stand sufficiently near a speaker to hear what is spoken the means whereby the sounds of his voice are carried to our ears are as follows: as he speaks, his vocal apparatus produces what are called *sound-waves;* i. e., to-and-fro motions in the air. The rapidity of these to-and-fro motions or their number in a given time, varies with the pitch of the sounds, and their loudness varies with the energy of the motions. When these sound waves enter the ear of the listener they produce to-and-fro motions of a membrane or drum head situated at the end of the alley-way leading into the ear, which vibrations, when transferred to the nerves of hearing, cause the sensation of sound.

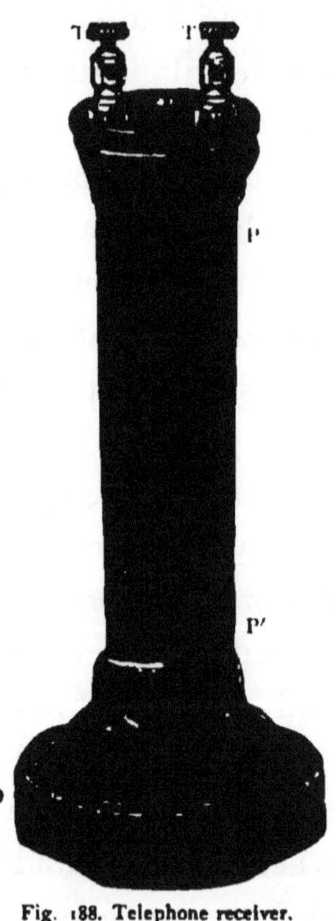

Fig. 188. Telephone receiver.

Let us now try to understand the manner in which the receiving apparatus reproduces the words which are spoken at the transmitting end of the line. We will take for this purpose the well known form of *telephone receiver* shown in Fig. 188, which is connected with the line terminals by the binding posts at T, T. Fig. 189, shows the construction of this receiver. It contains a permanent magnet M, placed in the tube P,P′, Fig. 188, and provided with a coil of insulated wire C, near its end P. This coil is provided with a soft iron core, connected to the same end of the magnet which can be

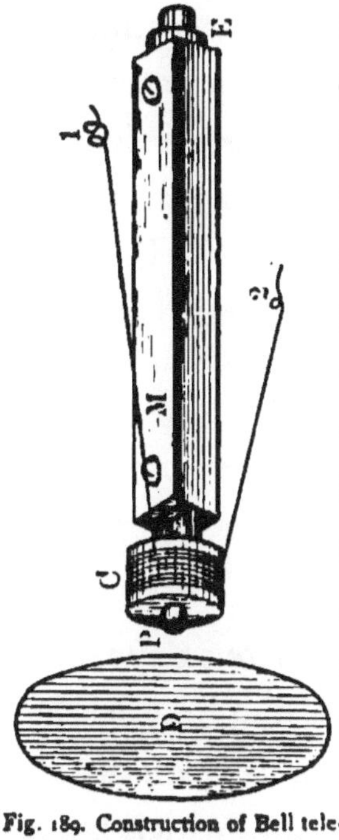

Fig. 189. Construction of Bell telephone receiver.

seen projecting beyond the coil at P. Directly in front of the coil is placed a *diaphragm* D, of thin iron plate, such as is used for ferrotypes. The terminals of the circuit are connected to wires or conductors 1 and 2. The magnet renders the soft iron core and the diaphragm D, magnetic, and magnetic flux passes from the core to the diaphragm, thus passing through the coil C.

Fig. 190. simple telephone circuit.

When the sound-waves from a speaker's voice strike against the diaphragm, they move it to-and-fro, and cause variations in the amount of magnetic flux which moves into and out from the coil, thus generating E. M. F's in it. If the transmitting instrument is connected by means of a circuit with a similar telephone receiver

T′, as shown in Fig. 190, these electric currents passing over the line will flow through the coil of wire on T′, in alternately opposite directions. When the current flows in a direction such as to strengthen the pull of the receiving magnet, it will attract its armature or diaphragm ; when its flows in the opposite direction, it will weaken the pull, thus permitting the diaphragm to move back under the influence of its elasticity. There will thus be reproduced in the diaphragm of the receiving instrument all the movements that were produced in the diaphragm of the transmitting instrument by the speaker's voice, so that any one listening at the receiver T′, will be able to hear what is spoken at T. Consequently, as we have already said, the speaker's voice furnishes the power to drive the transmitter T, as a dynamo and the currents flowing from the line, drive the receiver T′, as a motor and causes it to reproduce all that has been spoken into T.

It has been found preferable in practice to use a different form of transmitter than that used for the receiving instrument. There are a great variety of this kind of transmitters. They belong to what are called *microphone transmitters.* The *microphone* is a device invented by Prof. Hughes, whereby feeble sounds may be increased in intensity by means of a telephone placed in the circuit of a voltaic battery. The sound waves strike the microphone and cause it to vary the electric resistance of the telephone circuit and thus produce sounds in it. Fig. 191, shows a very simple form of microphone. The voltaic cell P, and telephone T, are connected in the circuit as shown, which contains a number of loose contacts formed by three nails at C, C′, C″. If any sound be made in the neighborhood of the nails, the sound-waves will alter their contact surfaces and thus vary the electrical resistance of the circuit with a rapidity which is exactly the rapidity with which the sound waves move to-and-fro. Consequently, any one listening at T, can hear what is said in the neighborhood of the nails.

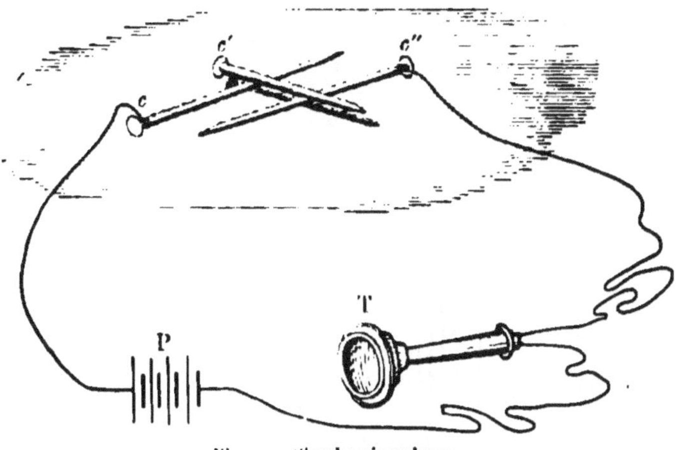

Fig. 191. Simple microphone.

Instead of causing the energy of the speaker's voice to drive a miniature dynamo, and thus produce varying E. M. F.'s, which reproduce the sounds in the receiving instrument, it may be employed to vary the resistance of a circuit containing a voltaic cell, the varied currents so

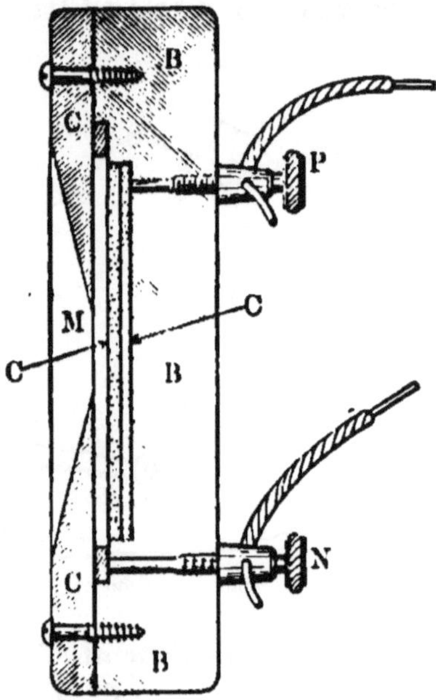

Fig. 192. Dust telephone transmitter.

set up reproducing the sounds in a telephone receiver like that shown in Fig. 189.

Various forms have been given to microphone transmitters. A form which gives excellent results and which is shown in Fig. 192, is called a *dust transmitter*. One of the circuit wires P and N, is connected to the plate of carbon C, and the other to an elastic plate P, of platinum foil, which acts as a diaphragm. The space between these two plates is filled with granulated carbon. An inspection of the figure will show that these plates, with their interposed layer of granulated carbon, are placed directly opposite the mouth piece M. As the diaphragm is moved to-and-fro by the sound-waves, the resistance of the carbon contacts is varied

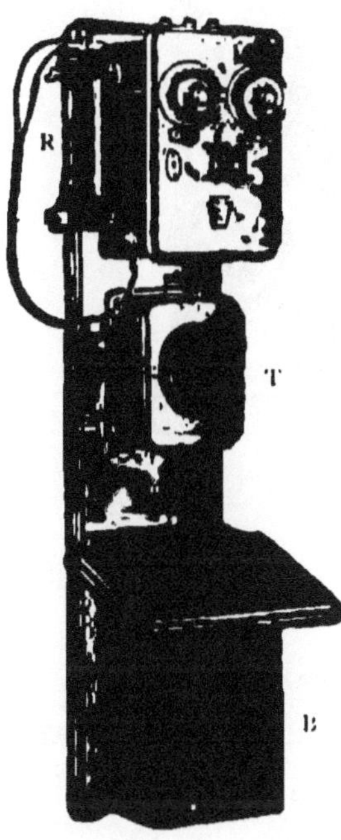

Fig. 193. Telephone apparatus.

in exact accordance with the to-and-fro motions, and so the current flows over the line with to-and-fro variation exactly corresponding to the to-and-fro motions of the sound-waves.

Fig. 194. Form of Telephone apparatus.

Fig. 193, shows the telephonic apparatus that is generally placed at each end of a line. T, is a microphone transmitter and R, the receiver. A voltaic battery is placed in the box B, for operating the transmitter. A call bell is shown at M. Another form of telephonic apparatus is shown in Fig. 194. The transmitter is shown at T, the receiver at R, and the bell at M.

CHAPTER XVII.

SOME OTHER APPLICATIONS OF ELECTRICITY.

WHEN the circuit of a voltaic battery of 3 or 4 cells is broken, only a very small spark is seen. If, however a coil containing many turns of insulated wire be placed in the circuit of the battery, on breaking the circuit, a bright spark will be seen, of a pressure sufficiently great to enable the discharge to jump across a short air space. This spark is produced as follows :—At the moment of breaking the circuit, the magnetic flux, which always accompanies a current, and which fills the loops of the coil while the current is flowing through it, rapidly dies out of the coil, thus setting up in it an E. M. F. of considerable strength.

The electric spark so produced may be

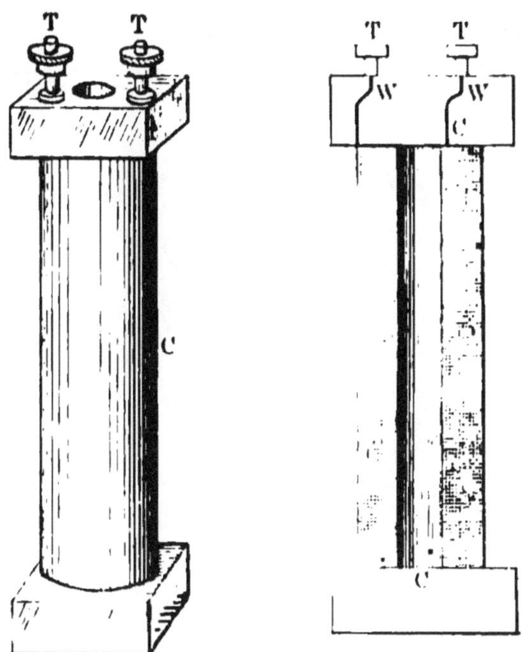

Fig. 195. Elevation and section of spark coils.

used for the ignition of gas jets. For this purpose a coil, called a *spark coil*, is employed, consisting of many turns of insulated wire wrapped around an iron core,

Such a coil is shown in Fig. 195. The spark coil is placed in the circuit of a few voltaic cells and the gas burner that is to be lighted.

Fig. 196, shows a *pendant gas burner*

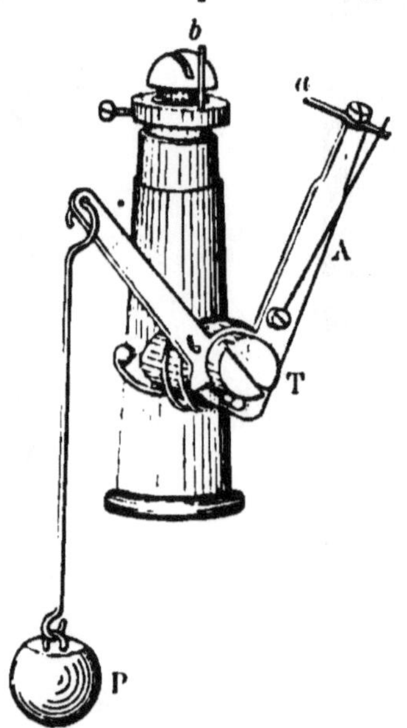

Fig. 196. Pendent gas burner.

employed for causing the sparks from a spark coil to ignite a gas jet. This burner is furnished with a pendant P, the pulling of which turns on the gas by turning a key at T, brings the lever A, so that the contact wire b, comes into contact with another contact wire a, thus closing the circuit at a and b. On releasing the

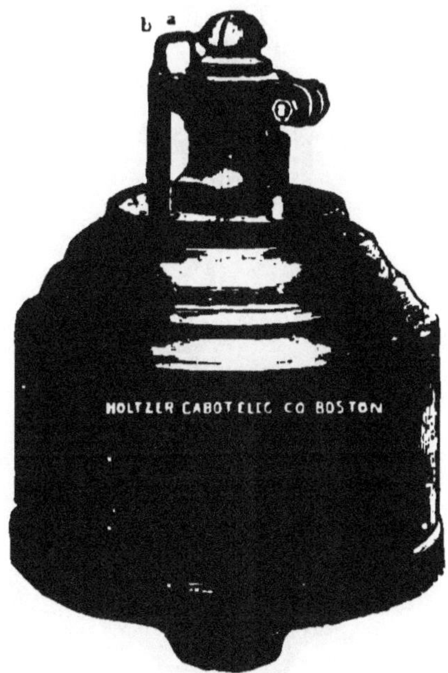

Fig. 197. Automatic burner.

pendant, the lever flies back, thus breaking the circuit and forming a spark in the issuing stream of gas, which thus lights it. When it is desired to extinguish the light a second pulling of the pendant P, turns off the gas by the key T.

Sometimes a device called an *automatic burner* is employed. Here the touching

Fig. 198. Interior mechanism of automatic burner.

of one button, usually a white one, turns on and lights the gas, and the touching of another button, usually a black one, turns off the gas. A form of automatic gas-burner is shown in Figs. 197 and 198. It contains two electro-magnets M, and M', one placed in the circuit of the white button, for the turning on of the gas, and the other, in the circuit of the black button, for turning it off. At the same time that the gas is turned on, a form of automatic contact breaker keeps up a constant vibration of the contact points a and b, with a stream of sparks between them, and this is maintained as long as the push button keeps the circuit closed. The advantage of an automatic burner lies in the fact that by its use the gas can be readily lighted and extinguished from a distance. The method of operating such a burner is shown in Fig. 199.

Another method for electrically lighting an issuing gas jet is shown in Fig. 200. This apparatus, called a *portable gas-lighter*, consists of a small spark coil and a battery, the circuit of which is opened

Fig. 199. Automatic burner.

and closed by means of a key operated by the hand.

The electro-magnet finds a number of applications in houses and hotels for various calls, alarms, etc., such for example as for fire or burglar alarms, or for calls generally.

Fig. 200. Portable gas burner.

In a *hotel annunciator* each of the different rooms is connected by a separate circuit with an electro-magnet placed in the annunciator. On closing the circuit in any room, the attraction of the arma-

Fig. 201. Gravity-drop annunciator.

ture of the particular electro-magnet connected with that circuit causes a shutter to drop displaying the number or name of the room calling. At the same time an electric bell is rung, thus calling the attention of the attendant in the office to the fact that some room desires a service. In the form of *gravity-drop annunciator* shown in Fig. 201, the drops have fallen at the bath room, hall and parlor showing that these rooms have called for some service.

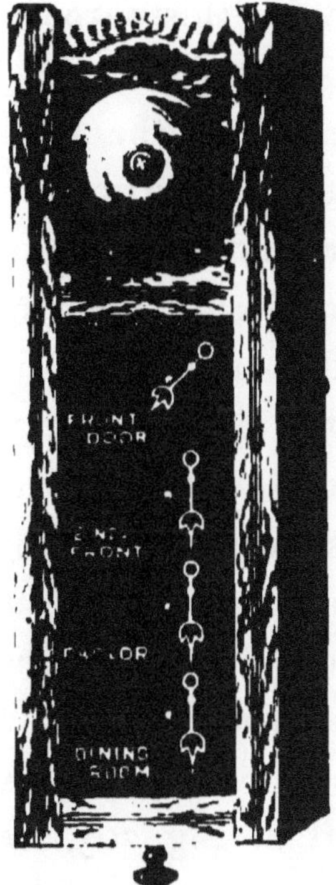

Fig. 202. House annunciator.

Instead of causing the attractions of the armature of the magnet to drop a shutter, it may move a needle, thus showing the circuit calling. A form of *needle annunciator* is shown in Fig. 202. Here the needle shows that the front door is calling.

In all annunciators, some device is provided for resetting the drops or needles after they have been moved. In Fig. 202, a push is provided for this purpose, at P.

Burglar alarms operate on the principle of the annunciator. Electric contacts are placed on doors and windows, or on the stairs, so that on closing them an alarm is sounded, and a needle on the annunciator indicates the exact point at which the contact has been closed. In the form of burglar alarm shown in Fig. 203 the attached clock is so arranged that the alarm is cut out of the circuit when the hands reach a certain predetermined position, at a certain time. The switches shown in the lower part of Fig. 203, are for cutting out the alarm from any particular circuit by hand.

Electricity has been applied in a variety of ways to *fire alarms.* A simple plan

Fig. 203. Burglar alarm with clock cut-out and other attachments.

consists in means whereby, in some cases, a device called a *thermostat* is so placed in the circuit of an electro-magnetic bell that when a certain increase of temperature

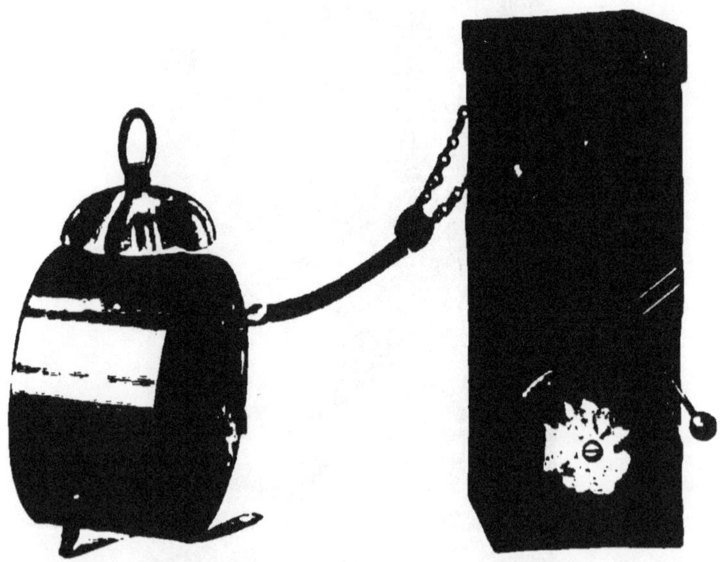

Fig. 204. Cut shows circuit connected up to alarm clock ready for duty.

has been reached the circuit is closed and the bell sounds an alarm.

An electric bell may be used in connection with the alarm clock shown in Fig. 204, for the purpose of calling a sleeper at a given time. In this case the connections are such that as soon as a certain hour is reached the clock closes an electric circuit, in which are placed a voltaic battery and bell, which then sounds an alarm.

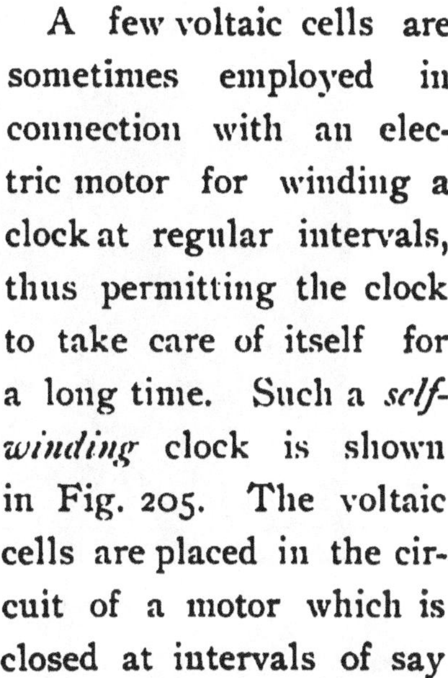

A few voltaic cells are sometimes employed in connection with an electric motor for winding a clock at regular intervals, thus permitting the clock to take care of itself for a long time. Such a *self-winding* clock is shown in Fig. 205. The voltaic cells are placed in the circuit of a motor which is closed at intervals of say every half hour, by contacts operated by the works. The battery power required for this purpose is very small, so

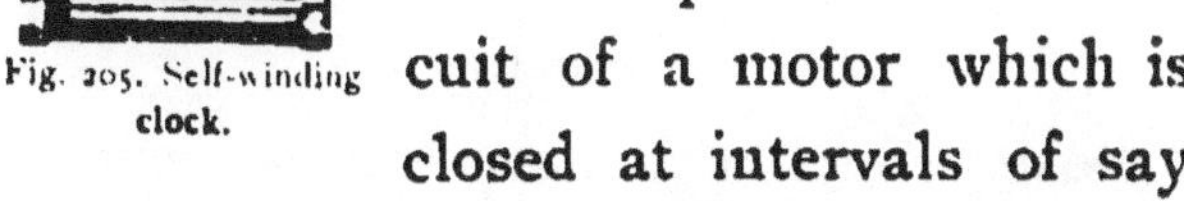

Fig. 205. Self-winding clock.

that a small battery will attend to the winding of a clock for a long time.

Miniature electric lamps are sometimes employed as electric jewelry. As we have seen, small lamps require only a few volts pressure to operate them, so that a very small battery will suffice for this

Fig. 206. Electric Jewelry.

purpose. Such a battery can be readily carried in the coat pocket. When it is desired to light the lamp, the circuit is closed by means of a key as shown in Fig. 206.

An important piece of electric apparatus known as a *call box* is shown in Fig. 206. By pulling down the handle until it reaches the different calls marked on the box, and then releasing it, a wheel is set in motion which sends an interrupted current over a telegraph line extending to a central station, thus indicating from the character of the impulses received, both the house or place calling and the nature of the service required.

Fig. 207. Call box.

While, as we have seen, electricity is a servant willing to administer to our comforts and interests, yet it is at times a hard master. There are two dangers which are connected with the use of electricity; namely, the danger to property from fire, and the danger to person from shock. The danger from fire can only occur when the wires carry an unduly strong current or, when they are imperfect. The danger to the person can only occur with wires of high pressure.

We have already pointed out how the danger arising from the overheating of a circuit by an abnormally strong current is avoided by the use of the safety catch or fuse. It is for this reason that where electric wires are properly installed the fire danger is practically removed. Considering the amount of power which an electric circuit is often capable of transmitting, it is wonderful with what safety to property electric circuits can be installed.

It is a fortunate circumstance that the electric resistance of the human body is so high that danger to life can only occur from high electric pressures. There is practically no danger to life from powerful electric currents unaccompanied by high electric pressures. The worst that a powerful electric current can do is to heat a wire which carries it, and such a wire may be incautiously grasped by the hand and so cause a burn. A high pressure, however, is of course able to overcome the resistance offered by the body, and to send a sufficiently powerful current through the body to result even in instant death. Care should be observed, therefore, in handling wires which may be directly or indirectly connected to a high pressure. An incandescent lighting circuit, employing only 110 volts, or, perhaps, 220 volts, may be regarded as fairly safe. A street-car circuit of 500 volts is capable of killing horses, and should be handled with caution. Arc-light circuits should never be handled by inexperienced people, since the pressure employed on such circuits is commonly thousands of volts.

Generally, however, it is preferable that those not acquainted with electricity should avoid handling any circuit wires, since even though a wire may belong to a system in which only a low voltage is employed, yet it may accidentally happen to be in electrical contact with a circuit operated by a dangerously high voltage. Should occasion arise, however, for handling a live wire, or a circuit which may be regarded as dangerous, it is well to remember that if the body be insulated, such a wire may be safely grasped with one hand, since the electric current requires not only a contact at the place where it enters the body, but also a contact at the place where it leaves it, in order to send a current through the body;

otherwise, no complete circuit can be formed through the body and no bad effects can be produced. At the same time, however, it should not be forgotten that if a circuit be touched by one hand, while accidentally in contact with the ground at some other portion of its course, the person is practically touching another part of that circuit through the ground with his feet, a position which is, probably, the best he could take to receive a reverse shock.

THE END.

INDEX.

B.

C.

D.

E.

F.

G.

H.

I.

M.

N.

* 9 7 8 3 3 3 7 3 9 3 9 2 2 *